Risk Analysis Strategies for Credible and Defensible Utility Decisions

Awwa
Research
Foundation

About the Awwa Research Foundation

The Awwa Research Foundation (AwwaRF) is a member-supported, international, nonprofit organization that sponsors research to enable water utilities, public health agencies, and other professionals to provide safe and affordable drinking water to consumers.

The Foundation's mission is to advance the science of water to improve the quality of life. To achieve this mission, the Foundation sponsors studies on all aspects of drinking water, including supply and resources, treatment, monitoring and analysis, distribution, management, and health effects. Funding for research is provided primarily by subscription payments from approximately 1,000 utilities, consulting firms, and manufacturers in North America and abroad. Additional funding comes from collaborative partnerships with other national and international organizations, allowing for resources to be leveraged, expertise to be shared, and broad-based knowledge to be developed and disseminated. Government funding serves as a third source of research dollars.

From its headquarters in Denver, Colorado, the Foundation's staff directs and supports the efforts of more than 800 volunteers who serve on the board of trustees and various committees. These volunteers represent many facets of the water industry, and contribute their expertise to select and monitor research studies that benefit the entire drinking water community.

The results of research are disseminated through a number of channels, including reports, the Web site, conferences, and periodicals.

For subscribers, the Foundation serves as a cooperative program in which water suppliers unite to pool their resources. By applying Foundation research findings, these water suppliers can save substantial costs and stay on the leading edge of drinking water science and technology. Since its inception, AwwaRF has supplied the water community with more than $300 million in applied research.

More information about the Foundation and how to become a subscriber is available on the Web at **www.awwarf.org**.

Risk Analysis Strategies for Credible and Defensible Utility Decisions

Prepared by:
Simon Pollard
Centre for Water Science, School of Applied Sciences, Cranfield University
Bedfordshire, MK43 OAL, UK

Steve Hrudey
School of Public Health, University of Alberta, Edmonton, AB, Canada T6G 2R3

Paul Hamilton, **Brian MacGillivray**, **John Strutt**, **John Sharp**, and **Roland Bradshaw**
School of Applied Sciences, Cranfield University, Bedfordshire, MK43 OAL, UK

William Leiss
McLaughlin Center for Population Health Risk Assessment, 1 Stewart Street,
University of Ottawa, ON, Canada K1N 6N5

and

Alan Godfree
United Utilities Water plc, Public Health Section, Lingley Mere Business Park,
Great Sankey, Warrington, WA5 3LP, UK

Sponsored by:
Awwa Research Foundation
6666 West Quincy Avenue
Denver, CO 80235-3098

Published by:

Awwa Research Foundation

American Water Works Association

IWA Publishing

DISCLAIMER

This study was funded by the Awwa Research Foundation (AwwaRF). AwwaRF assumes no responsibility for the content of the research study reported in this publication or for the opinions or statements of fact expressed in the report. The mention of trade names for commercial products does not represent or imply the approval or endorsement of AwwaRF. This report is presented solely for informational purposes.

CONTENTS

TABLES

FIGURES

FOREWORD

The Awwa Research Foundation (AwwaRF) is a nonprofit corporation that is dedicated to the implementation of a research effort to help utilities respond to regulatory requirements and traditional high-priority concerns of the industry. The research agenda is developed through a process of consultation with subscribers and drinking water professionals. Under the umbrella of a Strategic Research Plan, the Research Advisory Council prioritizes the suggested projects based upon current and future needs, applicability, and past work; the recommendations are forwarded to the Board of Trustees for final selection. The foundation also sponsors research projects through an unsolicited proposal process; the Collaborative Research, Research Applications, and Tailored Collaboration programs; and various joint research efforts with organizations such as the U.S. Environmental Protection Agency, the U.S. Bureau of Reclamation, and the Association of California Water Agencies.

This publication is a result of one of these sponsored studies, and it is hoped that its findings will be applied in communities throughout the world. The following report serves not only as a means of communicating the results of the water industry's centralized research program but also as a tool to enlist the further support of the nonmember utilities and individuals.

Projects are managed closely from their inception to the final report by the foundation's staff and large cadre of volunteers who willingly contribute their time and expertise. The foundation serves a planning and management function and awards contracts to other institutions such as water utilities, universities, and engineering firms. The funding for this research effort comes primarily from the Subscription Program, through which water utilities subscribe to the research program and make an annual payment proportionate to the volume of water they deliver and consultants and manufacturers subscribe based on their annual billings. The program offers a cost-effective and fair method for funding research in the public interest.

A broad spectrum of water supply issues is addressed by the foundation's research agenda: resources, treatment and operations, distribution and storage, water quality and analysis, toxicology, economics, and management. The ultimate purpose of the coordinated effort is to assist water suppliers to provide the highest possible quality of water economically and reliably. The true benefits are realized when the results are implemented at the utility level. The foundation's trustees are pleased to offer this publication as a contribution toward that end.

Walter J. Bishop
Chair, Board of Trustees
Awwa Research Foundation

Robert C. Renner, P.E.
Executive Director
Awwa Research Foundation

FOREWORD

ACKNOWLEDGMENTS

The authors are indebted to the following water utilities, organizations, and individuals for their cooperation and willing participation in this research:

- Awwa Research Foundation, Denver, Colo.: Rick Karlin, Linda Reekie, and Bob Allen
- Canadian Water Network, Waterloo, Ont.: Mark Servos and Ed McBean
- City of Calgary Waterworks, Calgary, Alta.: Paul Fesko
- City of Toronto, Toronto, Ont.: John Rudnickas, Patrick Newland, and Deborah Blumentrath
- Cleveland Water Division, Cleveland, Ohio: Greg Wilson and Marlene Sundheimer
- Concerned Walkerton Citizens, Walkerton, Ont.: Bruce Davidson
- Cranfield University, Cranfield, Bedfordshire, U.K.: Tom Stephenson
- Environmental Science and Research, New Zealand: Chris Nokes
- EPCOR Water Services, Edmonton, Alta.: Allan Davies and Les Gammie
- Her Majesty's Treasury, London, U.K.: Chris Butler
- Hydro One, Toronto, Ont.: John Fraser
- Louisville Water Company, Louisville, Ky.: Bob Miller and Don McKay
- National Grid, Warwick, U.K.: Geoff Hobson and Gareth Llewellyn
- New York City Department of Environmental Protection, New York, N.Y.: David Lipsky, Michael Principe, and Christopher Ward
- Philadelphia Water Department, Philadelphia, Pa.: David Katz and Kumar Kishinchand
- Scottish Water, Dunfermline, Scotland: Gareth Lifton
- Severn Trent Water, Birmingham, U.K.: Nicola Content
- South Australia Department of Health, Adelaide, South Australia: David Cunliffe
- South Australian Water Corporation, Adelaide, South Australia: Don Bursill, John Howard, Karen Simpson, and Annelie Lourens
- South Gippsland Water, Victoria, New South Wales: Brian Ashworth
- Thames Water Utilities, Reading, U.K.: David Holt, Fiona Davidge, and Tony Rachwal
- United Kingdom Water Industry Research, London, U.K.: Michael Farrimond
- United Utilities Water, Warrington, U.K.: Geoff Miller and Steve Whipp
- University of Alberta, Edmonton, Alta.: Elizabeth Hrudey
- Veolia Water, Hatfield, U.K.: Richard Lake
- Water Corporation of Western Australia, Leederville, Western Australia: Richard Walker
- Yorkshire Water Services, Bradford, U.K.: Derek Wilson, Robert Lloyd, and Peter Abell

The authors are especially grateful to the Awwa Research Foundation (AwwaRF) Project Advisory Committee who willingly gave their professional time to support this research:

- Linda Reekie, project manager, AwwaRF, Denver, Colo.
- Tim Gablehouse, Colorado Emergency Planning Commission, Denver, Colo.
- Steve Estes-Smargiassi, Massachusetts Water Resources Authority, Boston, Mass.
- Greg Wilson, Cleveland Water Division, Cleveland, Ohio

Acknowledgment and thanks are also extended to Edna Holliday (Cranfield University) for her expert assistance in formatting and proofreading this report.

EXECUTIVE SUMMARY

A MESSAGE FOR UTILITY MANAGERS

The water utility industry's approach to risk management is increasingly explicit and integrated with other business processes. This formal adoption of risk management within the sector codifies many implicit practices and recognizes that the provision of safe drinking water should be a "high reliability" service within society, subject to the rigors and controls inherent to other sectors, such as the nuclear, offshore, and aerospace industries. These sectors have learned important lessons about implementing risk management that are transferable to the water utility sector.

This report is a critical review of risk analysis strategies used by water utilities for better decision making. It has been informed by an international workshop on risk analysis, several discussions with utility executives and risk managers, multiple presentations, and structured interviews with operational utility staff. The report summarizes the status of risk management in the water utility sector among the participating and collaborating utilities.

CONCLUSIONS

The following conclusions have been drawn from this research:

- A tendency remains to view risk assessment as an end rather than a means to inform company exposures and opportunities for innovation.
- The value of risk management requires greater advocacy at a sectoral level and within individual organizations.
- The promotion of water safety plans is driving an integrated approach to public health risk analysis from catchment (watershed) to tap. Implementation of a water safety plan approach is a powerful vehicle for a renewed emphasis on risk management within water utilities.
- Risks and values are intimately linked; however, there is little explicit expression, at present, of the risks that utilities are prepared to accept, whether the utilities have private, public, or corporatized governance structures.
- Failure to manage risk in a climate of efficiency gains may result in the sector becoming exposed to the types of organizational disaster witnessed during a fatal outbreak of waterborne illness in Walkerton, Ont. Regardless of its complexity, the tolerance of risk within an organization must be discussed and confronted.
- Securing a risk management culture on the ground within a water utility requires effective risk knowledge management, tenacity in addressing the underlying causes of incidents, and strong leadership. Further research that addresses risk management cultures within the sector is warranted.
- The compelling case for effective risk management should be the core argument of water utilities for moving toward an organizational structure that can be more strategic and forward-thinking.

CHAPTER 1
INTRODUCTION AND OBJECTIVES

PROJECT NEED AND RATIONALE

In the twenty-first century, risk management has become the water utility manager's central function. Beyond the preserves of public health protection and the maintenance of process performance, utility managers must also assess and manage legal, regulatory, financial, environmental, resource and reputational risks in a complex and demanding business climate that is subject to an unprecedented changing pace (Means, Ospina, and Patrick 2005). The implications of their decisions extend well beyond company structures to public agencies, regulators, consumers, elected officials, politicians, and shareholders. In attendance is an increasing scrutiny of utility decisions. How then should the water sector address the risk management agenda in a practical and meaningful way?

This report recommends that utilities move swiftly to embed risk management within their organizational cultures as a step toward becoming more strategic and forward-thinking. This involves adopting a more mature organizational capability in risk management, one that ensures the provision of safe drinking water alongside a less risk-averse approach to asset management and technology adoption. The work presented here continues a stream of Awwa Research Foundation (AwwaRF)-funded research on risk management, water utility infrastructure, and organizational change (AwwaRF 1997, 2004, 2005, 2006a–e).

Although the techniques of risk identification, risk analysis, risk management, and risk communication have been applied within other process industries, their explicit and routine use within the water utility sector is relatively new. This is because the sector has relied on the successful implementation of quality standards to ensure that risks are managed. However, the current climate of change demands a fresh look at the challenges to the sector and the strategies it might adopt for better, more credible decision making within a future climate that is far less certain. Simply stated, compliance with quality standards does not guarantee that one operates an inherently safe system.

Considerable scope exists within this debate for learning from the experiences of related utility sectors and from "best in class" organizations. These combined experiences can prove invaluable to the water sector as it considers the risk analysis strategies and decision-making frameworks it might adopt for more defensible and rewarding utility decisions.

STUDY OBJECTIVES

The research has adopted three objectives:

1. Completion of a baseline assessment summarizing current risk management frameworks, techniques, case studies, and "best practice" examples, and requiring the following elements:
   - A review of risk analysis frameworks and their application within the utility sector
   - The elicitation of views from risk managers and policymakers across a range of sectors and management levels on implementation issues

   - A cross-sectoral benchmarking of risk analysis and decision-making capability among selected utilities
2. Collation of alternative approaches to managing risk that were organized as part of this research and delivered through an international workshop—Risk Analysis Strategies for Better and More Defensible Decision-Making.
3. Completion and analysis of case studies, which are presented in this report as examples of good and bad practice in risk management. The cases are supplemented by a detailed study of the risk management capabilities of two participating water utilities.

This report has been supplemented by a series of research papers, published conference proceedings, and related articles (Pollard et al. 2004, 2005, 2006a, b; MacGillivray et al. 2006a).

CHAPTER 2
BASELINE ASSESSMENT

CLASSES OF UTILITY DECISIONS

Risk management is the principal business of water utilities. With respect to formalized risk assessment and management, water utilities have performed risk assessments in a relatively restricted context, such as in addressing specific safety, health, and environmental risk concerns. More recently, however, risk techniques have been extended throughout organizations as utilities seek to inform decision making at all business levels (Figure 2.1), managing all the risks to which they are exposed.

Strategic Decisions

At the strategic level, corporate risk managers manage a plethora of financial, political, and regulatory risks associated with merger and acquisition activity, company reputation, outsourcing, and the long-term viability of investment decisions. Many companies now have risk committees in place to monitor, prioritize, and report on these risks to the board, thereby meeting the requirements of internal financial control.

Beyond the organization, risk managers need to consider risks external to their operations because water utilities are exposed to commercial and market risks (Rothstein and Kiyosaki 2003), including currency and interest rate fluctuations, competitive dynamics, and stock market volatility. In this context, recognizing and managing risk and opportunity becomes an essential feature of strategic risk management.

A key issue for utilities is to define their "appetite for risk." This is important because it defines a utility's lowest tolerable limit of risk and, by inference, the costs they are prepared to invest in risk management. Notwithstanding the approval of capital maintenance plans, for example, it appears that debates about risk appetite are rare within utilities. McKay (2005) relates how Louisville Water Company (Kentucky) has actively sought to be risk taking (i.e., opportunity seeking) in the areas of utility expansion and in the adoption of new technologies. This utility has assumed more risk in their insurance program, employing large self-insured retentions and deductibles, some of which are market driven.

Tactical (Program-Level) Decisions

Program-level decisions involve tactics—the transfer of strategy into action. Such decisions often concern the rollout of utilitywide activities such as staffing policies, asset management planning, and approaches to outsourcing.

An increasing role for risk management in the United States is in asset management. In 2001 the American Society of Civil Engineers (ASCE) reported that the U.S. drinking water system faced an annual shortfall of $11 billion for replacing facilities nearing the end of their useful lives, with sewer systems requiring similar levels of investment (ASCE 2001). The condition of underground assets is not easy to gauge and costs are high. Many companies are supporting preventive maintenance programs that include a range of monitoring techniques and risk tools for evaluating the condition and criticality of assets, which allows prioritization of

Source: Adapted from Strategy Unit 2002.

Figure 2.1 Risk hierarchy

capital spending (Stone et al. 2002, Allison and Chadwick 2003, Lifton and Smeaton 2003, Radovanovic and Marlin 2003). Similar program-level utility upgrades have been made in response to the recent security agenda (U.S. Government Accountability Office 2006).

Program-level risk analysis involves the prioritization of activity and risk–cost trade-offs. Well-designed, piloted, and continuously improved risk ranking systems can provide a basis for targeting risk management (Pollard et al. 2002). Whether used to drive capital investment priorities, maintenance schedules, monitoring regimes, or workforce planning, these systems may also *incur* risk unless the consequences of resource trade-offs are assessed in advance. Specifically, caution is advised against the use of risk ranking tools without cross-checking the implications for items that may not be funded, after having generated a list of prioritized expenditures.

Operational Decisions

Operational decisions concern implementation "on the ground" as well as routine and nonroutine operations. Increasingly, water utilities use a range of techniques (WHO [World Health Organization] 2004) to minimize the risk of microbial and chemical contamination in water supply systems. Although largely used to improve operations (e.g., better catchment management), these activities can have wide implications at the tactical (e.g., investment prioritization) and strategic levels (e.g., risk appetite), highlighting significant interfaces between these and other risk management activities within organizations (Table 2.1). These considerations are often contentious and challenging for discussion among organizations.

Successful implementation of these activities necessitates a shift in operational and regulatory focus from environmental technologies and contaminant targets toward management-based strategies that have a demonstrable benefit of risk reduction and product improvement—a more strategic approach to water utility management. International initiatives such as the *Australian Drinking Water Guidelines* (NHMRC [National Health and Medical Research Council] 2004), the WHO water safety plans (WHO 2004), and the Bonn Charter (IWA [International Water Association] 2004) are providing excellent incentives for such management. The current emphasis on regulating individual parameter limits through treated water compliance monitoring in isolation of the broader aspects of risk management inherent to regulatory schemes in the United States and

Table 2.1
Decisions for safe drinking water supplies

Decision level	Decision
Strategic	What is the utility's risk appetite in relation to its drinking water supplies?
	What priority should be given to water safety risks in relation to other risks (e.g., employee health and safety, environmental risks, financial risks)?
	What values are applied with regard to water supply risks? For example, are attempts made to give all customers the same quality of service or is spending prioritized on a risk-reduction-per-dollar basis?
Tactical	How much needs to be spent to balance risks throughout the organization?
	What is the most cost-effective way of achieving risk reduction?
	How much needs to be invested in maintaining the human and physical asset infrastructure?
Operational*	What hazards exist in the water supply systems and what are their associated risks?
	To what extent are the controls effective?
	How can significant risks be reduced or mitigated?

*These are essentially the objectives of the World Health Organization's water safety plans (WHO 2004).

Canada, however, offers little current incentive for pursuing a comprehensive risk management approach.

CRITERIA FOR CREDIBLE, DEFENSIBLE DECISIONS

Better utility decisions are not only those informed by risk analysis but those in which stakeholders have greater confidence. In a business climate where customer expectations are high and continue to rise (Patrick and Means 2005), utility decision making should be

- Credible in the sense that decisions are better informed and, where appropriate, subject to greater transparency and rigor
- Defensible in that companies are able to demonstrate they have operated with due diligence and within acceptable levels of risk

Evidence-based and risk-informed decision making are frequently cited as criteria for good decisions. Demonstrating such criteria may require the following elements:

- Adherence to industry codes and standards
- Compliance with risk acceptability criteria
- Demonstration, through supporting analyses and with quality data, that risks are being managed to the lowest levels reasonably practicable
- Application of cost–benefit analysis
- Prioritization of risk control measures on the basis of risk reduction per unit cost

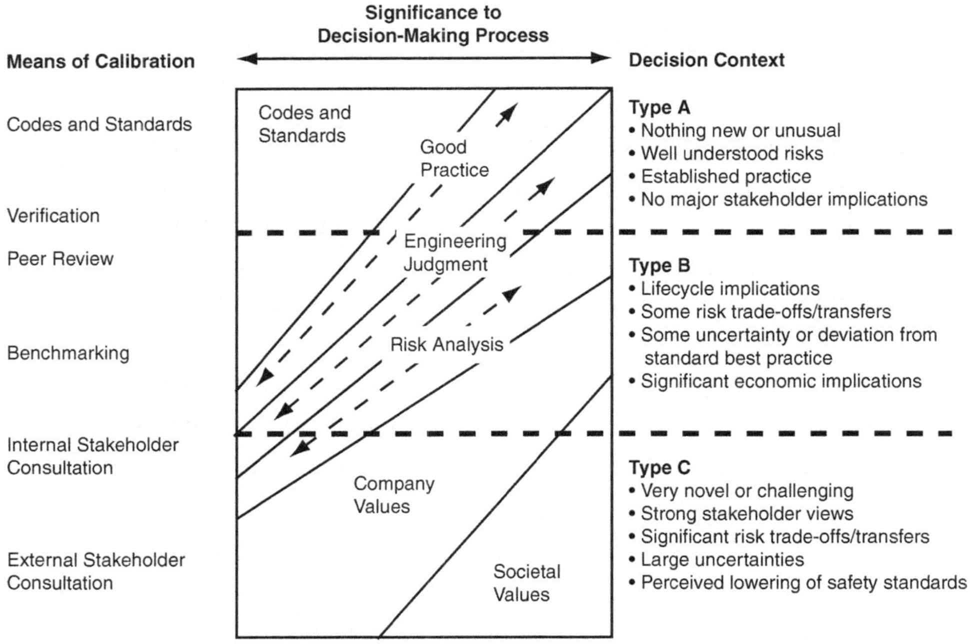

Source: UKOOA 1999, with kind permission from the U.K. Offshore Operators Association.

Figure 2.2 Decision framework

- Formal appraisals of the options for risk management, including an analysis of the potential unexpected outcomes
- Adherence to explicit company values
- Full consultation with stakeholder groups

The U.K. Offshore Operators Association (UKOOA 1999) has developed a decision-making framework that is widely accepted and used by organizations in the offshore oil and gas sector (Figure 2.2). The framework provides a structured and integrated approach that enables business, technical, and social factors to be considered and used to establish a sound basis for decision making. These procedures are largely generic and appear applicable to the water sector.

RISK DRIVERS

> *Whereas water utilities and regulators once shaped the marketplace, customers, owners and stockholders, and elected officials will define the new one. Customer expectations will continue to rise, and markets must respond favorably to new customer concerns. (Huber 2000)*

The reasons for the recent growth and interest in risk management strategies have been reviewed elsewhere (e.g., Westerhoff et al. 1998, Deloach 2000, Pollard and Carroll 2001, Pollard et al. 2004). Utility decisions have widespread implications for all stakeholders, including shareholders, regulators, and consumers. Consumers and other stakeholders are growing intolerant to the specific types of risk popularized by the media and environmental activist groups; this places important responsibilities on utilities to continue to develop their education and participation programs focused on understanding and valuing water. Therefore, a key issue is the capacity for water companies to analyze their risks and demonstrate to consumers and regulators that risks are being managed and reduced to an acceptable level.

Corporate Governance

Since the 1990s, companies have been required to demonstrate sound corporate governance and to identify and manage risks, thereby protecting shareholder value. "Corporate governance" has diverse meanings for different people (Marshall 2001). For many, it is the improvement of supervision and stewardship over shareholder value, although it has become interchangeable with concepts of corporate social responsibility. Whatever the merits of various interpretations, many financial institutions see improved standards of corporate governance as a means for increasing a company's creditworthiness (Pollard and Carroll 2001, Marshall 2001). Amatetti (1994) describes a situation where a privately owned water utility in the United States lost significant creditworthiness because it failed to consider the long-term financial ramifications of an internal policy decision to increase treatment capacity. According to the author, the utility had not compared the cost of reducing leakage from its distribution system with the cost and risk associated with maintaining the additional support capacity.

A summary of international corporate governance and risk management reports is provided in Table 2.2. In the United States, a prescriptive method for evaluating business risk was described in the Committee of Sponsoring Organizations of the Treadway Commission (COSO) documents published in 1992 (COSO 1992), a process that has since become known as the control risk self-assessment (CRSA) approach. In 2004 COSO provided a comprehensive framework for enterprise risk management (ERM) to supersede the previous CRSA approach.

In the United Kingdom, the Turnbull report (ICAEW [Institute of Chartered Accountants in England and Wales] 1999a, b) requires that listed companies implement a sound system of internal control to safeguard shareholders' investments and companies' assets. Under Turnbull, directors are charged with the annual reporting to shareholders of business risks and mitigation measures. Recent emphasis has been on increasing stakeholder awareness of company objectives, corporate risks, and changing organizational behaviors. A generic risk register for water utilities has been adapted for this report (Table 2.3). Case study 1 describes how Louisville Water Company uses risk inventories to inform board-level decisions on managing business risk.

It is proposed that effective risk management cannot be achieved without an effective, goal-oriented management structure. Many recommendations discussed in this report most effectively apply to utilities organized as business units, irrespective of whether they are publicly or privately owned. In contrast, the most common situation in North America, for example, is where water utilities are run as departments within the local (municipal or regional) government, which may be less conducive to these types of approaches. The compelling case for achieving effective risk management should be used by water utilities as a core argument for moving toward an organizational structure that can be more strategic. Case studies 1, 2, and 3 illustrate how a strategic view of business risk is being implemented in three utilities within a sector.

This move toward a more strategic approach is supported by the Swain report, recently commissioned by the Ontario provincial government to provide advice on the organization and long-term financing of Ontario's water and wastewater systems (Swain, Lazard, and Pine 2005). The report argues that a wide range of changes are needed for the challenges ahead, suggesting that systems must "join together to better manage their risks," inferring a capacity for enterprise-wide risk management.

Table 2.2
Summary of key reports and standards about corporate governance and risk management

Report	Country	Reference	Key points
AS/NZS [Australian/New Zealand Standard] 4360:1995, 1999, and 2004 *(Risk Management)*	Australia/New Zealand	AS/NZS (1995, 1999, 2004)	Provides generic guidance on risk management.
Where Were the Directors? Guidelines for Improved Corporate Governance in Canada (known as the Dey report)	Canada	Toronto Stock Exchange (1994)	Recommends that company boards assume responsibility for company stewardship, including risk management.
CAN/CSA-Q850-97, *Risk Management: Guideline for Decision-Makers*	Canada	Canadian Standards Association (CSA 1997)	Provides generic guidance on risk management.
Report on Corporate Governance, 1999: Five Years to the Dey	Canada	Toronto Stock Exchange (1999)	Identifies consistent areas of noncompliance with corporate governance guidelines.
Cadbury and Greenbury report (forerunner of *The Combined Code*)	United Kingdom	Committee on Corporate Governance (1998)	Codifies the practice for proving the sound financial probity of a company.
Turnbull report	United Kingdom	Institute of Chartered Accountants in England and Wales (ICAEW 1999a, b)	Recommends that organizations establish a risk-based approach to establish a system of internal control.
Internal Control—Integrated Framework	United States	Committee of Sponsoring Organizations of the Treadway Commission (COSO 1992)	Provides recommendations for management about how to evaluate, report, and improve control systems.
Enterprise Risk Management—Integrated Framework	United States	(COSO 2004)	Provides enterprise risk management framework that encompasses internal control, superseding previous guidance.

Table 2.3
Generic risk register for businesses

Generic risk register for businesses
Business
Wrong business strategy
Competitive pressure on price/market share
General economic problems
Regional economic problems
Political risks
Obsolescence of technology
Substitute products
Adverse government policy
Industry sector in decline
Take-over target
Inability to obtain further capital
Bad acquisition
Financial
Liquidity risk
Market risk
"Going concern" problems
Overtrading
Credit risk
Interest risk
Currency risk
High cost of capital
Treasury risk
Misuse of financial resources
Occurrence of fraud
Misstatement risk relating to published financial information
Breakdown of accounting system
Unrecorded liabilities
Unreliable accounting records
Penetration and attack of IT [information technology] systems by hackers
Decisions based on incomplete or faulty information
Too much data and not enough analysis
Unfulfilled promises to investors
Breach of listing rules
Breach of financial regulations
Litigation risk
Breach of competition laws
Breach of other regulators and laws
Tax penalties
Other
Business processes not aligned to strategic goals
Failure of major change initiative
Loss of entrepreneurial spirit
Stock out of raw materials
Skills shortage
Physical disasters (including fire and explosion)
Failure to create and exploit intangible assets
Other (continued)
Loss of intangible assets
Breach of confidentiality
Loss of physical assets
Lack of business continuity
Succession problems
Loss of key people
Inability to reduce cost base
Major customers impose tough contract obligations
Over-reliance on key suppliers or customers
Failure of new products or services
Failure to satisfy customers
Lack of orders
Failure of major project
Loss of key contracts
Inability to make use of the Internet
Failure to deliver by outsource provider
Industrial action
Failure of big technology-related project
Lack of employee motivation or efficiency
Inability to implement change
Inefficient/ineffective processing of documents
Poor brand management
Product liability
Inefficient/ineffective management process
Problems arising from exploiting employees in developing countries
Other business probity issues
Other issues giving rise to reputational problems
Missed business opportunity
Water industry specific
Contaminated or noncompliant water supply
Excessive leakage from distribution system
Mains burst
Corrosion of distribution system
Water shortage
Loss of water supply
Injury to staff or contractors
River pollution
Illegal connection
Sabotage
Asset failure
Odor from wastewater treatment plant
Noise pollution
Broken discharge consent
Pollution of watercourse
Flooding of customer property
Chlorine gas leak
Reservoir failure

Source: Adapted from ICAEW 1999a.

Case Study 1
Louisville (Kentucky) Water Company

One facet of the Louisville Water Company's risk management approach, adopted from the COSO model, is an assessment of so-called soft controls—the ethics and integrity embedded within a company's culture, operating standards, employee competencies, and empowerment (McKay 2005). Two questionnaires are used by internal audit to certify employee compliance with their code-of-conduct and code-of-ethics policies and to survey the effectiveness of internal controls. Results reflect the employees' evaluations of the "tone at the top," which is the environment of ethics, values, and risk tolerance boundaries established by management. Each of Louisville Water Company's twelve operating business systems annually identifies, assesses, and ranks the operating risks within its own system in facilitated brainstorming sessions and reviews them periodically. Most business systems conduct the sessions with their system advisory councils, which are comprised of both frontline and supervisory employees. Nothing is "off the table." Although risk inventories often include some frivolous ideas, McKay (2005) reflects that the sessions always introduce "real nuggets" that had not previously been considered. The rankings are generally qualitative and based on prescribed value scales.

Case Study 2
United Utilities Water plc (United Kingdom)

Miller (2005) describes an impressive program of initiatives taken by United Utilities to improve its risk reporting and monitoring capabilities. This includes establishment of the following components:

- Regular risk management meetings to review controls and actions on significant exposures
- A risk and issues database with established scoring protocols
- Procedures for escalating significant new risks to directors
- Appointed roles of risk champion and risk coordinator in each directorate. Risk champions are senior managers with the experience to test, challenge, and improve the efficiency and effectiveness of the risk management process. Risk coordinators implement the risk management process.

Case Study 3
Severn Trent Water (United Kingdom)

Similar to United Utilities, many of the larger U.K. water companies now operate central databases whereby potentially significant business risks can be identified by employees. Leverett (2003) details how Severn Trent Water continuously assesses routine and nonroutine tasks in five categories. All assessments have to be authorized at an appropriate level corresponding to the level of risk, with authorization that includes any risks to be tolerated or actions proposed. Significant risks are linked to the company's risk management database, allowing information to be shared throughout the company and the central tracking of actions.

The report further recommends that assets should be in the possession of and administered by a municipally owned corporation, and stresses the need for "strong and effective governance."

Providing Good, Safe Drinking Water

As well as protecting their businesses directly, many utilities use preventive risk-based strategies to protect their internal and external stakeholders (customers, employees, regulators, etc.) because of the many high-profile water incidents in the last two decades (Table 2.4). One such event was the 1998 *Cryptosporidium* crisis in Sydney, New South Wales (McClellan 1998, Cunliffe 2003, Hrudey and Hrudey 2004). A boil-water notice was issued after a high concentration of *Giardia* cysts and *Cryptosporidium* oocysts were found in raw and treated waters. Fortunately, the initial detections were false positives or the pathogens were inactive because no illness in the community resulted.

This incident was not without substantive costs to the utility (Table 2.4) and served as a wakeup call for the industry in terms of catchment management and the ways risks are communicated to the public. For commentators Deere and Davison (1998) and Cunliffe (2003), the incident highlighted the limitations of management based purely on end-product monitoring and the lack of a coordinated approach to managing water quality. The Sydney incident provided fertile ground for developing the *Framework for the Management of Drinking Water Quality* that was used to restructure the *Australian Drinking Water Guidelines* (NHMRC 2004).

In 2000 another incident took place in Walkerton, Ont. where the water supply became contaminated with microbial pathogens after manure was spread on a farm near a public supply well. More than 2,300 people became ill and 7 people died. The public inquiry (O'Connor 2002a) reported a litany of bad practices before and during the incident. These included the failure of operational staff to maintain effective chlorination of the water supply and perform routine chlorine residual checks, the falsification of water quality records by staff, and the failure of treatment managers to respond satisfactorily to positive microbial tests, any one of which may have significantly reduced the impact of the incident or prevented it altogether. The second report of the Walkerton Inquiry (O'Connor 2002b) recommended that water suppliers adopt a total quality management system based on preventive rather than strictly reactive strategies to identify and manage risks to public health.

Financial Pressures

More than ever, utilities must seek to balance spending with effective risk management. This has long been an implicit component of asset management. A recent report to the U.S. Senate (U.S. General Accounting Office 2004) explicitly endorses the role of risk analysis in targeting spending on the nation's water infrastructure. Within the United Kingdom, pressure from the economic regulator, the Office of Water Services (Ofwat), has ensured the incorporation of risk analysis into asset management programs. Water utilities are expected to "demonstrate how the flow of services to customers can be maintained at least-cost in terms of both capital maintenance and operating expenditure, recognising the trade-off between cost and risk, whilst ensuring compliance with statutory duties" (U.K. Ofwat 2000).

In addition to regulatory pressures, the global trend toward requiring financial self-sufficiency on the part of public and private utilities has created a climate in which utilities can no

Table 2.4
Selected water supply/water quality incidents

Incident	Location	Year	Cause	Cost to utility	Reference
Microbial contamination of drinking water*	Milwaukee, Wis. (United States)	1993	Suboptimum design and operation of a water treatment utility was suspected to have been responsible for a massive outbreak of cryptosporidiosis, believed to have affected about 400,000 people and left 69 dead.		Lisle and Rose (1995); Solo-Gabriele and Neumeister (1996)
	Sydney, New South Wales (Australia)	1998	High concentrations of *Cryptosporidium* oocysts and *Giardia* cysts were found in a utility's raw and treated water supply. No incidence of disease was reported, although a boil-water notice was issued to residents.	The cost to taxpayers was "tens of millions of pounds." The incident also caused the resignations of a chairperson, managing director, and two senior managers.	McClellan (1998)
	Walkerton, Ont. (Canada)	2000	Approximately 2,300 residents became ill and 7 died from drinking water contaminated with *Escherichia coli* O157:H7 and *Campylobacter jejuni*. The contamination was traced to the runoff from a local farm, which had infiltrated a nearby well. Chlorine residuals were not maintained, and operators failed to perform simple checks and falsified records.		O'Connor et al. (2002a); Hrudey et al. (2002)
Chemical contamination of drinking water	Camelford (United Kingdom)	1988	A stand-in delivery tanker driver pumped 20 tons of aluminum sulfate directly into the treated water reservoir of an unstaffed utility.		Environmental Data Services (1999); Buckley (2004)
	Glasgow, Scotland (United Kingdom)	1997	Approximately 200 L of diesel fuel was spilled during an unstaffed fuel transfer to a mobile generator. Unknown to the treatment operator, the drain into which this spillage had entered fed directly into the wash-water recovery system. The problem went unnoticed for 24 hours. In total, 60,000 customers were affected. The water was deemed unfit to drink for 8 days.	Four months after the incident, West of Scotland Water estimated costs arising from the incident to be US$6.5 million (£3.3 million). The chairperson also resigned.	Fraser (1998)
Unplanned drought conditions	Yorkshire (United Kingdom)	1995	The water company had to transport up to 700 tankers per day (71.5 ML/d) of drinking water for several months.	Tanker transportation cost a total of US$92.5 million (£47 million). The government considered removal of the company's operating license.	Uff (1996)
Reservoir incidents	Los Angeles, Calif. (United States)	1963	The failure of the Baldwin Hills reservoir led to the loss of the entire water supply reservoir. A neighboring residential area was devastated by a raging flood. Despite extensive evacuations, 5 people died. About 41 homes were destroyed and 1,000 damaged.		Jansen (1988)
Major power cut	Cleveland, Ohio (United States)	2003	A massive power outage in the northeastern United States and Ontario (Canada) disrupted water supplies for nearly 6 million people in Cleveland, Ohio, and Detroit, Mich. In Cleveland, approximately three fourths of the city's 1.5 million residents were completely without water for about 12 hours.	Cleveland Water Division is currently considering an expenditure of greater than $20 million to purchase backup power generators.	Cascos et al. (2004); American Water Works Association (AWWA 2003)

*Other tabulated summaries of disease outbreaks are presented in Hrudey et al. (2002), Lisle and Rose (1995), Solo-Gabriele and Neumeister (1996), and Badenoch (1990).

Case Study 4
North Battleford (Saskatchewan) Cryptosporidiosis Outbreak

Faced with budget cuts in the mid-1990s, the Saskatchewan Department of Environment and Resource Management (SERM), which held legislative responsibility for the drinking water program in Saskatchewan, drastically reduced their already-limited field inspection and enforcement of municipal drinking water facilities. This culminated in a SERM proposal to eliminate its drinking water program altogether. The proposal was tentatively approved by the Treasury Board in 2000–2001 and justified as being based on risk-based decision making. The subsequent North Battleford cryptosporidiosis outbreak, infecting between 5,800 and 7,100 persons in the immediate community plus a large number of visitors from three other provinces, led to a public inquiry into the outbreak and the provincial drinking water regulatory system.

Justice Laing (2002) concluded in the Inquiry report "that the current risk-based model employed by SERM since 1996 is arrived at on the basis of economics, and has nothing to do with how best to safeguard the health of the population, all of whom consume water."

longer overengineer facilities with the presumption of managing process risk through excessive and redundant margins of safety.

The tensions that arise between those seeking economic efficiencies and the preservation of the principal goal of providing safe drinking water are often apparent in the conflicting expectations and presumed purposes of risk analysis made by different professionals (case study 4). Some commentators recognize the consequences of stripping away margins of safety, redundancy, and protection in situations where "risk analysis" is misused to justify such actions. These tactics may render the system, as a whole, to be less safe, more precarious, and susceptible to catastrophic failure. In these situations, optimization programs, risk-centered maintenance scheduling, and risk-based monitoring schemes require careful scrutiny as to whether the purported cost reductions allow for the unintended consequences of implementing these programs.

WHAT IS RISK MANAGEMENT?

"Risk management" is a practitioner-led discipline. Its implementation within organizations can vary significantly according to their size, operations, ownership, and leadership, as well as the political, economic, legislative, and regulatory climate in which organizations operate (case study 5). Likewise, the usage of risk terminology can vary greatly. Unless otherwise stated, the risk terminology used in this report is in accordance with definitions presented in the ISO/IEC (International Organization for Standardization/International Electrotechnical Commission) guidelines, which is an international standard vocabulary (ISO/IEC 2002).

Risk Assessments in the Water Sector

Risk assessments for drinking water originated in the United States in the early 1970s (Haas, Rose, and Gerba 1999). Amendments to the Safe Drinking Water Act in 1974 (USEPA 1996) demanded that better estimates of the consequences from hazards be found for risk management purposes. This was followed by a series of studies from the U.S. National Academy

Case Study 5
Functional Separation of Risk Assessment and Management

Within the context of water utility risk management, most risk-based decision making is conducted in-house. In some larger companies, routine risk assessment activities tend to be coordinated through specialist teams with risk managers responsible for communicating key risks and controls up to the executive board level, the role of the risk management being not to "own" risks themselves but inform decision makers (Miller 2005).

In other companies, especially smaller ones, where risk management may be conducted on a more ad hoc basis, this separation may be absent with risk management performed by the same individuals responsible for decision making and coordinating risk management activities, as shown in interviews:

> *We're not big enough to have separate groups. The whole organization is just over 60 people. I [Operations Manager] am responsible for initiating a lot of our risk management.... The risk register comes under the Financial Manager, who does the corporate tasks.*

of Sciences (NAS), formerly the National Research Council, including their 1983 "Red Book" (NAS 1983) that provided a basic set of steps for assessing risks to human health from chemicals, including during drinking water exposures. Individual risk assessments have been used to set drinking water guidelines or standards for specific parameters. Unfortunately, until very recently, the resulting tables of numbers have not been coordinated into an overall coherent risk management approach for ensuring drinking water safety.

The 1997 Presidential/Congressional Commission on Risk Assessment and Risk Management (1997) recognized that "the output of risk assessment…often seems too focused on refining assumption-laden mathematical estimates of small risks associated with exposure to individual chemicals rather than on the overall goal—risk reduction and improved health status." The commission set out a modern framework for risk-based decision making. It emphasized risk management as an iterative process and formally recognized the central role of stakeholders—that is, individuals, groups, or organizations who can affect or (perceive to) be affected by the risk early on in the assessment process.

The modern features of risk management are reinforced in a review of "failed decisions" (Leiss 2005). This review is critical of situations where risk assessment is seen as the *raison d'être* without sufficient attention to the development and implementation of appropriate risk management measures to protect the public. Leiss cites the 2003 Canadian bovine spongiform encephalopathy (BSE) crisis as an example of poor risk management, whereby the risks of BSE in cattle had been quantitatively assessed in terms of likelihood but not consequence. Leiss concludes that organizations should adhere to the following recommendations:

- Complete and continuously update risk profiles (in terms of likelihood and consequence) pertinent to the organization.
- Mandate continuous improvement in risk reduction and mitigation for all of the most highly ranked risks to be managed.
- Ensure stakeholder engagement as the real and vital center of the risk management enterprise.

Case Study 6

Honolulu (Hawaii) Board of Water Supply

Within the water industry, ERM is gaining favor. An early example of portfolio risk management in the water sector is presented by Rothstein and Kiyosaki (2003). Their approach, facilitated by tools that include multiattribute analysis and scenario planning, involves determining the set of investments available to the water utility and their characteristics in terms of risks (both monetary and nonmonetary), returns, and resource requirements; prioritizing investment opportunities throughout functional areas (i.e., looking beyond departmental barriers within organizations); and selecting investment options by virtue of their alignment with established corporate goals and objectives.

Case Study 7

Enterprise Risk Management at Hydro One (Toronto, Ont.)

Hydro One is a large, publicly owned electricity delivery company in Ontario. They embarked on an ERM program in 1999, which involved

- Appointment of a chief risk officer
- Development of an ERM policy, defining the governing principles and identifying individual responsibilities for different risk management activities
- Development of an ERM framework that established the basic process for all risk assessments and provided uniform processes to identify, measure, treat, and report on key risks

More details of this program are given in a case study by Aabo, Fraser, and Simkins (2004). According to Fraser (2005), the primary objectives of this initiative were to

- Align the degree of tolerance for risks between the board, management team, and staff
- Plan and allocate resources for asset management in a defensible and structured manner based on a risk prioritization methodology
- Ensure a common understanding of the risks to the business objectives to ensure these are addressed on a timely basis
- Improve governance by monitoring and reporting to the board about major residual risks to the organization

In most organizations, such improvements require development, promotion, and stewardship of a risk management culture (case studies 6 and 7).

Modern Utility Risk Management

Typically, risk management is no longer seen by organizations as an isolated activity informing a specific decision or set of decisions but as an integrated and continuous part of business activity. An important milestone in its development in this regard was Australian/New Zealand Standard (AS/NZS) 4360:1995, 1999, and 2004, the generic risk management standard.

Case Study 8
Yorkshire Water (United Kingdom)

Since 2001, Yorkshire Water has been developing a state-of-the-art asset risk scoring system called Leading Edge Asset Decisions Assessment (LEADA). This forms the basis of part B of the water service provider's approach to the Ofwat "common framework." In LEADA, Yorkshire Water claims to have successfully developed a unified process that allows the comparison of risk throughout the entire asset base, including both source-to-tap (water) and sink-to-river (wastewater) assets. Another feature of the system is that the customer's "willingness to pay" (for improvements through their utility bills, as determined from detailed customer focus groups and surveys) has also been built into the model (Lloyd and Abell 2005).

This highlighted the need for strong strategic planning in support of effective risk management in organizations. Critically, the standard recommended identification of the following components:

- Roles and responsibilities of various parts of the organization for managing risk
- Relationships between the project and other parts of the organization
- Goals, objectives, strategies, scope, and parameters for risk management

This reiterates the need for some utilities to move toward more business-oriented organizational structures to conduct more effective risk management.

ERM (case studies 6 and 7) refers to this integration of risk management into a company's business processes. Contrasted with defensive or control-oriented approaches used to manage downside risks, ERM optimizes business performance by supporting and influencing pricing, resource allocation, and other business decisions. This is being witnessed in the water utility sector with the integration of risk ranking systems used to inform capital improvement programs (case studies 8 and 9) within the broader strategic business framework (case study 9).

Deloach (2000) reflects on the lack of consistency among organizations in terms of the level of detail, reporting formats, management methods, and guidelines for risk management. He advocates an approach that

- Implements a common language, facilitating internal and external communication
- Provides a consistent reporting framework for aggregating risk measures and information
- Fosters increased management confidence through a systematic approach that identifies all of an enterprise's risks
- Supports resource allocation through rigorous prioritization of risk
- Reduces transfer costs because of offsetting or pooling of risk
- Creates a disciplined, structured process for making vital decisions

RISK MANAGEMENT FRAMEWORKS

Risk management frameworks are "route maps" that describe formalized procedures for risk-based decision making. Commonly represented in a diagrammatic form and outlining an

Case Study 9
Scottish Water (Scotland)

A slightly different system is Scottish Water's asset risk and criticality scoring mechanism, Customer/Asset Risk Evaluation System, which endeavors to assess the total business impact of asset failures across the company in relation to its core business objectives (Lifton and Smeaton 2003, 2005). This is a source-to-tap Failure Mode and Effects Criticality Analysis (FMECA) with the addition of a business impact scoring system. Such a system assigns points to asset failure scenarios according to criteria such as loss of service, environmental impact, and loss of reputation.

In this system, one point is approximately equal to US$1,969 (£1,000) of adverse business impact. This provides what the authors refer to as a "common currency of risk" that allows a consistent approach for risk scoring to be incorporated across the business. The result is an integrated system that extends throughout the program and operational levels of the business, with the same tools being used to prioritize capital investment, plan capital maintenance, and assess the likelihood of plant maintenance schedules, thereby minimizing the risk of electrical and mechanical failure. One advantage of this system is that it provides a greater resolution for risk assessment, permitting risks that would have appeared within the same band to be prioritized.

assess-plan-do-monitor-revise management process, they have been broadened to include all identification and implementation elements that an organization needs to effectively identify, assess, and manage its risks (Jardine et al. 2003). The AS/NZS 4360 (2004) defines "risk management frameworks" as "a set of elements of an organization management system concerned with managing risk," a definition adopted in this report. A review of risk management frameworks by category is provided in Table 2.5.

A key issue for the water industry is the extent to which water companies are using these frameworks in practice. As discussed later, frameworks of interest to risk managers interviewed as part of this project fell into two main groups: corporate level frameworks for risk management and those advocated by industry regulators (e.g., water safety plans).

Corporate Risk Management Frameworks

The AS/NZS 4360 risk management standard (1995, 1999, and 2004) is most widely applied in corporate risk management frameworks. It is relatively simple to use without adaptation and identifies the necessity for strong strategic planning to provide effective risk management in organizations. In 2004, the AS/NZS updated the 4360 standard, placing greater emphasis on embedding risk management practices within an organization's culture and processes, and the management of potential gains as well as losses.

As with other business organizations, water utilities are seeking to establish sound internal control processes and ERM practices throughout their organization. Frameworks such as the AS/NZS 4360, as well as the more-specific risk governance frameworks (ICAEW 1999a, b; COSO 2004), are suited to this. It is evident that the concepts of corporate governance and ERM are becoming linked in some cases. COSO's ERM framework has already encompassed and superseded its earlier internal control guidelines (COSO 2004). Likewise, PricewaterhouseCoopers (2004) has also taken an integrated view of governance, risk, and compliance (GRC) in

Table 2.5
Examples of frameworks available for utility risk management

Area of use	Framework	Reference
Risk assessment frameworks		
Ecological risk	Framework for ecological risk assessment	U.S. Environmental Protection Agency (USEPA 1992)
	Guidelines for ecological risk assessment	USEPA (1998)
Microbiological risk	Conceptual framework for human health (chemical) risk assessment	NAS (1983)
	Conceptual framework for assessing the risks of human disease after exposure to waterborne pathogens	International Life Sciences Institute (ILSI 1996)
	Revised framework for microbial risk assessment	ILSI (2000)
Public health risk	Risk assessment model	enHealth Council (2002)
Frameworks for risk-based decision making/risk management		
Public health risk	Framework for environmental health risk management	Presidential/Congressional Commission on Risk Assessment and Risk Management (1997)
General	CAN/CSA-Q850-97; Q850 Risk management decision making	Canadian Standards Association (CSA 1997)
	AS/NZS 4360:2004; Risk management overview	See AS/NZS (1995, 1999, 2004)
	AS/NZS 4360:2004; Risk management process	
	AS/NZS 4360:2004; Risk treatment process	
	Risk management process	Association of Insurance and Risk Managers, National Forum for Risk Management in the Public Sector, and Institute of Risk Management (AIRMIC, ALARM, and IRM 2002)
	Benchmark risk management framework	Network for Environmental Risk Assessment and Management (NERAM 2003)
	AS/NZS 4360:2004; Risk management process	AS/NZS (2004)
Regulatory prescribed frameworks for risk management		
Water safety	Framework for the management of drinking water quality	NHMRC (2001,2004)
	Public health risk management plan for drinking water supplies	New Zealand Ministry of Health (2001)
	Multibarrier approach	NERAM (2002)
	Water safety plan	U.K. Water Industry Research (UKWIR 2003)
	Guidelines for drinking water quality	WHO (2004)

(continued)

Table 2.5 (Continued)

Area of use	Framework	Reference
Capital maintenance planning	Common framework for capital maintenance	UKWIR (2002)
Security	Vulnerability assessment model	U.S. National Institute of Justice (USNIJ 2002)
Frameworks for corporate governance/ERM		
Corporate governance	Implementation of Turnbull	Institute of Chartered Accountants in England and Wales (ICAEW 1999a)
ERM	"Stepping stones" along the enterprise-wide risk management journey	Deloach (2000)
	"Launch pad" to enterprise-wide risk management	
	ERM framework	Lam (2003)
	ERM framework	COSO (2004)
	Integrated view of governance, ERM, and compliance	PricewaterhouseCoopers (2004)
	Four stages of ERM process	Lam (2005)

their GRC framework. Lam (2003) outlines an ERM framework that incorporates corporate governance along with the following elements:

- Line management (business strategy alignment)
- Portfolio management (think and act like a fund manager)
- Risk transfer (transfer out concentrated or inefficient risks)
- Risk analytics (develop advanced analytical tools)
- Data and technology resources (integrate data and system capabilities)

A few of these elements can be implemented without a discussion of company values and risk appetite (case study 10).

Regulatory-Prescribed Risk Management Frameworks

Water safety plans (e.g., UKWIR [U.K. Water Industry Research] 2003, WHO 2004) are frameworks that identify step-by-step processes for managing water contamination risks from catchment to tap. The underlying rationale is that delivering safe water is not only about achieving a high level of apparent compliance. Water supply systems can attain this and yet have latent flaws in their design or operation that are contributors to "accidents waiting to happen." These may include uncontrolled contamination hazards in catchments, flaws in treatment design, leaks in distribution systems, poor operator training, and bad operational practices.

Australia has driven much of the development of water safety plans. In 1996, the NHMRC revised its *Australian Drinking Water Guidelines* (see NHMRC 2004), thereby providing guidance on processes needed to achieve safety in addition to providing the guideline numbers. The

Case Study 10

Company Values and Trade-offs in Risk-Based Decision Making

A key issue reaffirmed by attendees at the Banff conference (Pollard et al. 2005) was the importance of explicitly stated company values in the risk-based decision-making process. For example, it was argued by some attendees that the consequence of a water quality incident in a large urban population was more serious than a similar incident in a small rural community and should be accounted for when planning capital maintenance. This was not popular with some managers and industry representatives who believe that all customers deserve the same water quality irrespective of demographics.

Another point of contention was the issue of whether public health risks can be expressed in economic terms and "traded off" against other risks in an ERM system. As a water service provider, some believed that public health risks were of such importance that they should be considered alone in absolute terms rather than as a tradable commodity. Other managers who expressed an opinion argued that, where finances are constrained, this is a legitimate way of balancing these alongside other important risks to which their stakeholders are exposed (e.g., health and safety, environmental, and financial risks) and maximizing service to their customers for every pound or dollar spent on assets.

An important observation is that even where risk trade-offs are not explicitly articulated, they may still be taking place. These hidden trade-offs that may come to light during or subsequent to incidents are under heavy scrutiny. We may find, in such circumstances, that they are then far less defensible than when the trade-offs were made.

Ultimately, it is up to water industry managers and their stakeholders to weigh these finely balanced moral judgments and decide which values and risk management objectives to adopt. The participants in this discussion were from utilities with relatively high capabilities in risk management who have faced these dilemmas as a direct result of using risk information in their decision making.

fact that testing does not guarantee the safety of water supplies made the case for maintaining "effective barriers to prevent contamination of the water supply system."

The Australian Cooperative Research Centre for Water Quality and Treatment developed the *Framework for Management of Drinking Water Quality* (NHMRC 2001). The framework listed a sequence of twelve elements considered to be "good practice" for the management of drinking water supplies from catchment to consumer. In 2002, the NHMRC provisionally redrafted the drinking water guidelines document (NHMRC 2002), incorporating these twelve elements as its focus. This major restructuring and rolling update of the *Australian Drinking Water Guidelines* (NHMRC 2004) was intended to introduce a standard approach throughout the industry and establish due diligence and credibility.

Regulatory frameworks for risk management are not limited to water safety issues. In the United States, under the 2002 Bioterrorism Act, drinking water utilities serving more than 3,300 persons are required to conduct a vulnerability assessment to mitigate the risk of serious consequences from vandalism, insider sabotage, or terrorist attack (USEPA [U.S. Environmental Protection Agency] 2005c). Several tools have been produced to aid this process, including a vulnerability assessment model produced by the U.S. National Institute of Justice (USNIJ 2002). This model describes a twelve-step procedure that compares relative risks and analyzes the effectiveness of existing controls. The assessment process is performed by an appointed project team led by a trained facilitator and is intended to deliver a series of recommendations for reducing the most serious risks. These may include improvements to physical protection (e.g., hardened doors, locks, electronic security systems), consequence reduction improvements (e.g., reduction or dispersion of stored chemicals), and process control protection measures (e.g., IT [information technology] security, backup power supply).

Under U.S. law, water and wastewater facilities must also submit a risk management plan (RMP) to the USEPA (2005b), with the intention of reducing risks for hazardous chemicals being released into the environment. By making RMPs available to the public, the USEPA hopes to stimulate communication between the industry and the public to improve safety. This is in accordance with the 1986 Emergency Planning and Community Right-to-Know Act (summarized in USEPA 2005a), which was passed in the United States after an incident in Bhopal, India (Kletz 2001).

At the program level, a risk analysis approach is also advocated by the U.K. financial regulator as a means to prioritize future capital maintenance (U.K. Ofwat 2000). In previous price reviews, Ofwat based its analysis on a historical assessment of companies' serviceability indicators and capital maintenance expenditures. A new approach developed over the course of these successive reviews in the form of a common framework for capital maintenance planning (UKWIR 2002, U.K. Ofwat 2003). This framework encourages water companies to adopt a risk-based approach to their capital maintenance plans, allowing Ofwat to consider future service forecasts as well as historic performance assessments. Spillett (2003) describes how previous performance review plans were very detailed (bottom-up) and, if fully implemented, would have resulted in huge bill increases. As a result, company demands for this additional financing were largely ignored. The risk-based approach in the common framework is intended to benefit the industry and regulators. It allows companies to prioritize future spending, providing a more credible case for the size of customers' water bills, and helps the regulator benchmark companies more accurately and respond to each company's individual claims in a more fair and transparent way. For the 2004 U.K. Ofwat periodic review, the majority of companies in England and Wales have adopted the framework in their business plans.

RISK ANALYSIS TOOLS AND TECHNIQUES

A key capability in utility risk management is the range and sophistication of risk analysis tools and techniques available at the strategic, tactical, and operational levels (Table 2.6), as reviewed by MacGillivray et al. (2006a).

Advocates of ERM such as Deloach (2000) and Lam (2003) point to the use of quantitative tools being indicative of a more sophisticated risk management program. This presupposes that these tools are used wisely and supported by suitable data, a requirement not often met. There is a distinct discrepancy between the academic literature and standard industry practice in regard to the application of these tools. MacGillivray et al. (2006a), for example, report this for the application of simulation and scenario analysis in the field of investment risk analysis. This may be explained, in part, by the observation that such techniques do not fit naturally into most companies' skill sets or capital-budgeting systems (Luehrman 1997).

Promising applications of quantitative risk analysis can be found in the literature, usually informing tactical decisions on prioritizing expenditure. These applications can be powerful when combined with geographic information systems (GISs; case study 11). Seattle Public Utilities (Washington), for example, adopted a risk-based approach to asset management, considering the likelihood and impact of pipe rupture with reference to age, material, location, and historical cost of repair (U.S. General Accounting Office 2004). Using this analysis, utility officials have been able to delineate their pipe network into areas of critical and noncritical risk and

Table 2.6
Examples of risk analysis tools and techniques for operational, tactical, and strategic risks in the water utility sector

Level	Example	Tools used	Reference
Strategic	Outsourcing	Scenario planning	Zsidisin, Panelli, and Upton (2000)
	Investment risk analysis—portfolio management for public utilities	Multiattribute analysis Scenario planning and analysis	Rothstein and Kiyosaki (2003)
	Financial risks of public and private partnerships for infrastructure projects	Simple, expected cost analysis Sensitivity analysis Monte Carlo simulation	Grimsey and Lewis (2002)
Tactical	Risk-based asset management	GIS-based infrastructure risk models	Doyle and Grabinsky (2003) Booth and Rogers (2001)
	Process risk assessment to help formulate asset management strategy	Risk ranking Failure mode and effects analysis (FMEA)	Abell and Askey (2001)
	Prioritizing expenditure on mains rehabilitation	Three-tiered risk assessment: 1. Risk ranking 2. Generic quantitative risk assessment model 3. Monitoring program with tailored quantitative risk assessment	Radovanovic and Marlin (2003)
Operational	Assessing reservoir safety—operations and structural condition	Fault and event-tree analysis	Parr and Cullen (1988)
	Health and safety hazard assessment—chemical use at water and wastewater treatment works	Risk ranking Hazard and operability (HAZOP) studies What-if checklists FMEA Fault and event-tree analysis	Wirth and Siebert (2000)
	Process risk assessment and project contingency planning	Risk ranking	Leverett (2003)

Case Study 11
GIS-based Assessment Tools

GIS technologies now play a critical role in asset management. At their most basic levels, GISs allow utilities to convert data ordinarily displayed on paper maps into an accessible digital format, representing an excellent method for storing and collating data for future use (Foster and McDonald 2000).

The level of detail (i.e., the layers of spatial data) contained within such systems varies widely. Kaufman and Wurtz (1998) describe the evolution of a GIS for a small utility in Michigan. An extensive inventory of asset condition records and failure and maintenance data is collated within the system, supporting the risk-based planning of capital improvement and maintenance works. Pertinently, the system was developed in only 3 months and cost less than $3,000.

Similarly, Booth and Rogers (2001) illustrated how the implementation of GIS technologies within an asset management decision support system can allow for the visual tracking of infrastructure assets and their associated risk factors.

Although applications of GIS technologies in support of asset management have proven to be powerful risk-tracking, visualization, and communication tools, they rarely utilize the capabilities of a GIS to spatially analyze data in the classical sense. Doyle and Grabinsky (2003) illustrate these capabilities through quantitatively relating Toronto's infrastructure deterioration to spatially variable corrosion risk factors, providing a basis for the identification of network areas most at risk from external corrosion. Such an approach may allow utility managers to better focus rehabilitation efforts because of a more complete understanding of the causative factors behind water mains deterioration.

allocate maintenance resources accordingly. Through adopting this approach, officials believe they are using staff resources more effectively and that the program, over time, will lead to a reduction in maintenance costs.

Kent, Treece, and Davies (2003) similarly describe the prioritization of investment strategies for trunk main maintenance at Dwr Cymru/Welsh Water (Wales). The methodology is based on the available records of asset performance, condition, and serviceability, which are stored on the company's water asset management database. StruMap (Northgate Information Solutions plc, United Kingdom), a software-based mapping system, allows clustered failures to be considered separately from random bursts, a task performed because the former are likely considered as being representative of systemic susceptibilities.

Qualitative tools may also be valuable at the program level where data may be limited, or when used as a first step in a longer "tiered" risk assessment process. Radovanovic and Marlin (2003) describe the prioritization of expenditure within Sydney Water's (New South Wales) program of water mains rehabilitation. This comprised three assessment tiers, increasing in complexity and cost:

1. Risk screening using a simple likelihood–consequence risk ranking process
2. A quantitative risk analysis using a generic probabilistic model
3. A more focused program of site-specific monitoring and data analysis

Case Study 12
Setting and Using Risk Acceptance Criteria

One participating utility held regular risk "challenge" meetings to review their current risk profile. The risk manager reflected on their difficulties in setting and interpreting risk acceptance criteria and their desire to adopt a flexible approach to decision making:

> *The pitfall we could fall into would be to believe that we can just put something into a system and it tells us the answer...Agreeing the thresholds for investment is an interesting thing to have to go through, shall we say because some are easy to set, if you're looking at performance measures, but some are extremely difficult. When you're looking at health and safety, for example, and you're looking at potentially injuring your own employees, it becomes a very emotive subject, and very difficult to set a threshold on our current severity scale...Perhaps the threshold for health and safety should be much lower? But when you look at the extent and cost of it, you have to start thinking about balancing it against all the other risks and problems. So setting thresholds is very difficult. But what it means is that you have to be quite flexible...It [challenge process] gives us that opportunity to be flexible, because it's not just about stabilising the red area [high risks], because if you've got something in the amber that could go into the red, then perhaps you need to stop it [from] getting there.*

The utility also maintains a strong view on reputation risk management:

> *The one thing that we don't have in our severity scale is anything on public relations or reputation. We chose to leave that out...because it tends to be secondary. We think that if we actually get the service right in the first place then the bad PR [public relations] should go away...It's about getting to the cause of the problem and tackling it, not just a symptom of it.*

At the operational level (where emphasis is typically placed on reducing risks to human health and safety, and the environment), it is clear that the most popular and most effective risk assessment tools used by water companies are also qualitative or semiquantitative in origin (such as the risk ranking systems used by Deere et al. 2001, Leverett 2003, and Lifton and Smeaton 2003). This may be partially due to their simplicity, ease of use, and relevance for many common problems.

Interpreting Risk Assessment Outputs

Notwithstanding their utility, caution must always be applied in interpreting risk assessments, especially quantitative risk estimates (case study 12). In addition to the inevitable issue of data quality that applies to all modeled systems (garbage in, garbage out), quantitative risk analysis (QRA) is usually meaningful only for narrowly defined (closed) rather than broad-scope, complex (open) issues. QRA is less suited to low-probability, high-consequence events, such as a dam failure, where probability data is severely limited. QRA can be misleading for health risk assessment, particularly for contentious issues such as carcinogens. Leiss (2005) refers to a "fetishism of numbers" in this regard and suggests that use of value judgments better reflects the position of utility risk managers and stakeholders' perceptions of risk:

> *Every risk manager needs to be fully aware that here he or she is in what may be called a "permanently contestable zone." In other words, both what is or should be a matter of*

Table 2.7
Command-and-control versus systems thinking

Business context	Command and control	Systems
Perspective	Top-down hierarchy, functional procedures	Outside-in process and flow
Attitude to customers	Contractual	What matters?
Decision making	Separated from work	Integrated with work
Measurement	Output, targets, standards: related to budget	Capability, variation: related to purpose
Attitude to suppliers	Contractual	Cooperative
Ethos	Control	Learning

Source: Adapted from Seddon 2000.

> *"concern" to anyone, and what a "reasonable" response—especially by someone in a position of responsibility—is to that concern, are always disputable. Only by keeping this elementary fact always uppermost in mind can risk managers hope to succeed in their difficult endeavour.*

Walker (2005) similarly argues that although a risk-based asset decision tool may be satisfactory for determining mains replacement programs, reservations should perhaps be held on the use of what could be viewed as a "black box" for life-and-death water quality issues:

> *In my experience, there is a tendency for people to "hide" behind these numbers as the "answer," rather than understand they are a tool to help with the exercise of professional judgement.... The trouble is pathogens aren't as predictable as water mains. When a system has inadequate barriers, I refuse to say the probability of personal harm is "rare." As far as I'm concerned, it could be tomorrow, and the day after and... This, together with the costs of a water quality incident...means water quality improvements must proceed as soon as practical. Which is just what the Executive decided by subjective judgement!*

Walker (2005) insists that a water supplier's focus should be on proactively seeking out threats to their water supplies, assessing them, and responding accordingly, not merely being content with good compliance figures.

On a similar theme, Seddon (2000) is critical of the control-and-command management style that pervades much of modern industry. This style begins with the separation of decision making from actual performance, thereby defining management's role as that of decision making. Seddon argues that this can lead to managers failing to understand how and why their organizations perform as they do. The preoccupation of management's belief that improving numbers equates to improving performance is a notion Seddon deems plausible but false. Instead, he promotes a "systems thinking" ethos where there is greater focus on stakeholder interest, organizational learning, and an appreciation of variability of risk (Table 2.7).

CHAPTER 3
VIEWS FROM THE INTERNATIONAL WATER SECTOR

STRUCTURED INTERVIEWS—OBJECTIVES AND METHODOLOGY

This research summary is based on a series of structured interviews with international utility risk managers and industry spokespeople. Eighteen interviews were conducted between May 13, 2004, and Aug. 2, 2005, with specialists from five countries (Australia, Canada, New Zealand, the United Kingdom, and the United States). Sixteen interviewees worked primarily within the water sector, and two worked for an energy utility. The structured interviews had three objectives: to provide a practical perspective of risk management throughout the water industry that would be reflected in the risk analysis reviews, to inform the development of a risk management capability maturity model, and to highlight key issues for discussion at an international workshop.

Questions (appendix A) were deliberately open-ended, allowing interviewees to offer opinions and experiences of different risk management strategies. Followup questions encouraged more detail and comment on specific initiatives and risk analysis strategies. Anonymous samples of the conversations are included in this chapter. They illustrate some of the critical implementation issues that emerge and reflect the range and balance of comments obtained.

CRITICAL IMPLEMENTATION ISSUES

Development of Risk Management

One of the most interesting points to emerge was the existence of different risk management drivers for water utilities and the pathways of development these organizations had taken. For example, utilities with an interest in risk-based asset management have developed strong capabilities at the tactical level to prioritize capital investment and maintenance spending. In other companies, the emphasis has been placed more on developing (operational) strategies for business protection, project delivery, or water safety:

> *Following the…incident, the director came to us and asked us what we could do to prevent any similar thing [from] happening again.*

> *I suppose it was driven by corporate governance legislation…and I think also pressure particularly from insurers and other stakeholders.*

> *Because we were growing so fast, we knew we had some plant upgrades down the road, and so we started to look at our drinking water quality, taking a source-to-tap view, and that led us into thinking, "How do you prioritise what you want to look at?" And that led us into thinking more about risk.*

> *That [the NHMRC framework] was our initial thinking, but then we went into the more formalised Enviro System (ISO 14001). Now we're moving towards an asset management organization. I'm not sure what's going to be the methodology…that's still an open area.*

Risk is such a big thing now. I mean it's carried out at all levels of the organization...In terms of formal structures there is a Risk Group that reports to the Board on "showstopper" risks (financial and operational). There is then another group, a cross-functional group which is all about asset management and programme risk.

For us there's been quite a change in the last five years in terms of moving from compliance to understanding risk. We've always had pretty good sampling results. The real watershed in our organization was the '96 Australian Drinking Water Guidelines...Prior to that the main emphasis had been on sampling and compliance. The guidance themselves made a strong play about barriers and things like that...We've got 900 tanks across the state that we look after and we physically inspected the roofs on all of those tanks to see if they were waterproof and...all 900 leaked. We went and looked at the chlorinators and found that there was a duty chlorinator but not a standby chlorinator. The guys had to work out when the cylinder was going to run out and ... didn't always get it right...So that was a real eye-opener for the organization...Then coming on top of that was Walkerton. Sydney precipitated the catchment stuff but really it was Walkerton that precipitated the understanding that complacency really was a major issue. We had bad things like they had at Walkerton...Supplying safe water was all about making sure you had things in place that were reliable and were going to do the job...We might get good sampling results but it's not the sampling results that make the water safe...You can get away with bad practice for a long time but eventually your pigeons will come home to roost...After that we put together a major capital programme and started from there.

We first got really interested in risk through looking at managing our catchments.

When I started my career, the concept of risk analysis was not well known in our organization. There were numerous claims and cost overruns on capital projects because our engineers weren't performing environmental risk analyses in the planning stage of projects. So, during the construction phase, we were paying for hauling away contaminated soils; or the project would be shut down because the contractor hadn't been advised of subsurface conditions. Sometimes we hadn't identified asbestos when contractors were ready for demolition work—just a lot of unforeseen field conditions...We saw a need for a more organized risk management program. When we were developing a strategic business plan,...and not just a plan for capital projects, but a plan that looked at strategic initiatives across the organization...we asked ourselves what we needed to do to make wiser investments in not just our infrastructure, but also in our human resources. That's when we decided to make risk analysis part of the way we do business. It was through the strategic planning effort that we identified a need for a Risk Management Group focused specifically on driving risk management principles and practices throughout the organization.

Integration of Risk Management Practices

When asked how utilities could improve their risk management capability, many identified a need to achieve better integration of risk management throughout their organization:

We have a standard severity scale which allows us to compare any asset in a system with any other asset in the system. So we actually compare risks straight across the board now, within clean water and likewise on the wastewater side...And where we're trying to move towards is trying to balance and calibrate between the two...so that we can switch resources either from clean to waste or vice versa...Aligning the scales is something we've spent quite a lot of work on...Is a customer having sewage in his property worse than going without clean water for 24 hours?

There's this thing about people starting "to talk a common language," but it's actually very important.

One of the challenges for any organization is in trying to integrate the understanding of risk management across a population of users when they're all using discrete tools which perhaps have different terminology, different scoring systems and different ways of presenting the outputs...I see my role as trying to take an integrative approach, as one where we don't necessarily have everybody using the same techniques, that patently isn't appropriate, but to ensure that people have a shared understanding of our approach to risk management, and wherever possible, that there is a read-across or an ability to interpret results from the output of one tool in the wider context of programme or strategic risk.

Initially, when the...database went live, anyone had access to it. We found that people had pet concerns and would tend to overestimate the risks, through not taking an objective view. They were then assigning actions against people unbeknownst to those people. So what we did then was to try to formalise it such that every department has a coordinator to ensure consistency.

The biggest problem with risk assessments that we've done is getting consistency...When we did the sanitary surveys, we put together the methodology but then we left it to the local people (within the organization) to do their own sort of assessment. Now that we look back in hindsight, there were a lot of inconsistencies...Some things that people didn't think were a problem, really were. As it turned out I don't think we got a very good picture from that exercise...It was partly objectivity but it was also partly lack of appreciation...Like if Farmer Jones's cows have always been drinking out of the streams and it's never caused a problem before, then they don't even think of it as a risk.

Linking it all together...At the moment we've identified a whole raft of things to be done. What we need is a programme that prioritises them on the basis of risk reduction—what will give us most "bangs to the buck"...When the improvements have been completed, what will we have gotten out of them? How much will it have dented on our corporate risk profile? What will be the residual risk afterwards?

Use of Risk Frameworks

Interviews revealed considerable discrepancy among practitioners about what constitutes a risk management framework (case study 13). Many interviewees referred to the likelihood–severity ranking systems (typically 5×5 matrices or similar) used to assess their companies' risks

Case Study 13

In-house Utility Frameworks for Risk-based Decision Making

Several water companies develop and use their own frameworks to standardize risk-based decision making throughout their organizations. One such company is Yorkshire Water (United Kingdom).

Yorkshire Water's standard approach is used within the company to identify and assess the risks of individual asset failures and particular projects. Although such frameworks are more commonly found in other utility sectors (offshore oil, energy supply, etc.), their use in the water sector is relatively rare, apart from those used in some good-practice examples.

A similar framework is used by the gas and electricity supplier National Grid (Warwick, United Kingdom). An impressive feature of their approach is that not only has the company developed this high-level framework, establishing a consistent approach to risk management throughout the company, but it has also produced different sets of accompanying documentation, containing more specific, low-level information necessary for its implementation throughout different departments, including procurement and logistics, construction, and HS&E (health, safety, and environment).

A different risk management framework from Hydro One, a large Canadian energy utility, is presented by Aabo, Fraser, and Simkins (2004). A key feature of their framework is the provision to communicate and consult with stakeholders at every level of the risk management process. In this respect, it mirrors the CSA-Q850-97 risk management standard for decision makers (CSA [Canadian Standards Association] 1997).

rather than any strategic route map or formal approach to the analysis of risk and appraisal of risk management options. This may be indicative of formalized procedures for risk assessment and reporting being employed without a corresponding understanding of how risk data should be used within a wider context of risk-based decision making.

Aside from these matrices and the AS/NZS 4360 framework, interviewees showed little working knowledge of, or enthusiasm for, the risk frameworks or standards in the field of safety, health, and environmental risk. In some ways, this is not surprising in that many have been designed primarily for administrators and policy makers rather than industry. Two types of framework generated higher levels of interest—those relating to risk governance and water safety plans:

> *At the moment, we have a bottom-up approach where each business unit is registered [to ISO14001] on its own...We're trying to figure out a way to develop a more corporate approach, so that the whole city would have an overall risk framework to use...As long as we're moving towards an asset management organization, we just have to get a lot more sophisticated in terms of standardising processes and how we give it a corporate view.*

> *We're trying to get closer to the COSO model...because it focuses a great deal on control evaluation and that is what we want to look at...You can have a risk that's indigenous to one system and they can manage it very well, but some of the controls they apply inhibit the ability of other areas to do business. So we're looking at the interaction between controls as well as the risks.*

> *We have a definite enterprise risk framework that we've established ourselves. It's pretty much formalised. We try to use the same tools consistently and the same reporting consistently.*

Lots of the concepts [relating to water safety plans] have been in existence for a long time, particularly in larger urban areas. Here there's nothing particularly new apart from that it's nice to get it all documented and having things clearly spelt out. The bigger benefits for us is more in extending those principles into the smaller country areas which previously may have been a bit "out of sight: out of mind"...We're having to cut the cloth a little bit to suit, so that at the smaller localities we're not doing all the bells and whistles, but you go through a process whereby you decide what the real major hazard is that this supply is confronting and what we need to do to make sure we've got it under control...The documentation is really just a process that we're wrapping it all up in, but getting the mind-set of the operators to understand these things is what really brings the bacon home, so to speak.

I guess there's no problem with these frameworks [NHMRC, AS/NZS 4360] at a high level and we drew from these, but they didn't go into enough detail for what we needed for our work. —The viewpoint from a small water utility representative

We looked at the two watersheds that supply our city water. We did a risk assessment of potential hazardous contamination events, looking at things like changes in land use, contaminated aquifer, forest fire, in terms of likelihood and impact. The source of that was the Australian framework [NHMRC]...In retrospect, it became a bit of a risk framework that helped us prioritise things and developed into a business plan.

Qualitative Versus Quantitative Risk Analysis

For the most part, qualitative tools form the mainstay of water companies' risk assessment programs. Opinion is divided on the utility of quantitative tools. Although some believed they could render their decisions to be more credible, others were more skeptical, citing the following issues:

- Perennial problems with data quantity and quality
- Lack of skills and resources to dedicate to complex tools (especially in smaller water utilities)
- The apparent ease in which numbers can be manipulated

One risk manager stressed the absolute importance of "understanding your business" and the value of subjective judgment in decision making. This interviewee suggested that some decision makers become preoccupied with risk data in a belief that it renders more credible decisions. This often coincides with a failure to adopt a systems view of their business:

Some of the difficulties...are that the [risk] assessments themselves are extremely subjective. So we always try to look for real performance evidence that we can measure to give more assurance that the scoring is credible.

People are very adept at manipulating numbers to give them the answers they want, to push their own projects forward.

Our greatest problem is with data quality. It's that old line, "You put garbage in, you get garbage out."

The subjectivity of it all is probably the biggest thing that people have to get to grips with...particularly as you go lower in the organization, I suppose, whether it's lack of ability, lack of confidence or lack of experience or whatever...I never have a problem exercising judgement in terms of decision making, but other people like to have numbers to make the decision for them...You've got to know your business. The people that don't know their business are the ones who want numbers. They think it makes it easier to make decisions... I don't think it is helpful, or it's only helpful if you know what's driving the numbers. People who control the purse strings, for example, like to have numbers so it's easier to make a decision but it doesn't necessarily make the decision better.

On a related theme, several risk managers expressed a distrust of computer-based risk assessment models and packages:

One thing that we've shied away from willfully has been software. To me, the software is nice and it organizes things, and so on, but it relieves people of their obligation to really analyze what they're doing...There are so many outside vendors of software for everything, and so far we've just had a bit of an aversion to those.

I've seen computer programs that will work out all your risks for you, but they were too slick. It was giving you the answers before you had thought through the questions.

Stakeholder Engagement

Although water companies tended to liaise closely with stakeholders on risk mitigation measures and welcomed constructive feedback on their risk assessment processes, it was discovered during the interviews that they were resistant to the idea of allowing stakeholders to participate in the risk assessment process. Different reasons were given, ranging from the need to preserve commercial confidentiality and concerns about the conflicting objectives of stakeholders to fears that stakeholder representatives would lack the specialized knowledge to participate positively in the risk assessment and thereby slow the process. Although this is to be expected, notwithstanding the academic debates on greater openness in risk assessment, it results in long days with respect to public participation in risk assessment within most organizations (case study 14). Interviewees also highlighted the difficulties they had experienced in communicating health and safety risks to the public, especially with media tendencies to over-sensationalize stories:

Truthfully, no...We have public engagement processes in place, but I can't say we use them for risk assessment.

Not if we can help it...There's an example that came up the other day. We'd been asked to work with the...[environmental regulator] on something and so we had this big debate on whether or not they should be in on the risk study or whether we should communicate it to them afterwards and seek their opinion...and we decided on the latter.

Case Study 14

Concerned Walkerton (Ontario) Citizens

Concerned Walkerton Citizens is a pressure group that was formed following an incident in Walkerton in 2000. In an interview, co-founder and current vice chair, Bruce Davidson, related his experiences in engaging with water companies on a range of public health and environmental issues, extracts of which are presented here:

> *I think the difficulty is that while it [public engagement] is a noble principle, the step of actually getting there is still one that is foreign to those institutional cultures, because they're not used to this kind of openness.*
>
> *What we're trying to do is open the door to greater communications; and even though we haven't been banging heads on hard issues, what it means is that I now have the ability to pick up the phone and speak directly to some of the top-level management of those companies and ask them questions, some of which may have fairly simple technical explanations, which the public may not be aware of. So what is starting to happen is that you are seeing a process by which you can sort out issues and hopefully open the door in a non-confrontational way, because the minute the backs go up on either side, some of the candour and some of the cutting through disappears.*
>
> *If we come to these meetings in crisis only and we've had no introduction, no familiarity, no level of comfort, then it's very, very difficult to take people at their word because they've had no track record of being honest with you and your impression is that the only time they ever come across with anything is when they were trying to downplay something...or say that the cause lies elsewhere. You don't have the sense so much that you're dealing with the PR department.*
>
> *Or quite often what people are used to is when something is virtually a done deal, then what will happen is that they'll bring forward a plan to the public and it's a question of saying, "We want you to say it's a good idea rather than starting at the ground level."*
>
> *Until you have a crisis, we're looking at a very small percentage of the public that are going to take a keen interest in it. There's going to be some key stakeholder groups and there's going to be I would suggest probably 3% of the public that would watch with any great interest under normal circumstances, but I think those people can be barometers of what the greater public would be concerned about.*

When asked whether there are any no-go areas for stakeholders' participation, Davidson responded:

> *I have to say generically that it is difficult for me to sit as a stakeholder in the public and give my approval not to know. What I want is a discussion of what those areas might be and see the rationale behind them.*
>
> *It's not a question of me necessarily agreeing with everything that the company would say or that the regulator would say, and clearly we would never reach that point but we have to be aware at least of where we can agree to disagree...But I think this idea of isolation...and that...we have to have the discussion on the terms of one of those parties only, does not breed respect, trust or transparency.*
>
> *If a company is stonewalling you all the way up and the regulators are looking at you as laypeople then you're setting a very bad template for communications at a time of crisis.*

You'd have to spend a long time explaining it to them [the stakeholders] and it would only slow the process down.

Possibly, but at the end of the day, it wouldn't be their money they [the regulators] would be spending. And so their objectives wouldn't be [the] same as ours.

Whatever the merits of these arguments, many in the water utility sectors are protective of their risk information and skeptical about allowing stakeholders to have far-reaching roles in risk-based decision making. This is not to say they are opposed to stakeholder participation in all areas of the business. Several interviewees pointed to the value of liaising with local residents, industries, and other land users to improve hazards management within catchments. This is especially important for small water suppliers. Others acknowledged they had consulted with interested parties on environmentally and socially sensitive projects (e.g., the siting of a wastewater treatment works) and liaised closely with regulators about sensitive compliance issues. Some interviewees described use of customer survey data in determining best business practice or when planning expenditure, thereby granting customers a passive role in decision making:

Where the stakeholders do influence the risk process…is on project risk. There we are very interested in the views of parish councils or local pressure groups as they can have a big impact on individual schemes. It can influence what we do in terms of solution.

Most of our risk assessments are carried out internally. Where these show significant risks we are keen to discuss them with the regulators… We want to know where we stand and to have a measured response.

I'm not sure about the issue of the regulators being involved. Certainly, I believe that…and this becomes more important, the smaller the system gets… that the more that you can draw on what local people know about what might influence the quality or otherwise of their source water, the better.

We've developed the process significantly since it started off…it's still based around a 5×5 matrix…but we can now bring in "customer willingness to pay" to see what problem brings the "best bangs for bucks."

The regulator influence is actually on the severity scale that we use.

The appetite of stakeholders for getting involved in the risk assessment process was identified as lacking. Some risk managers believed that regulators lacked sufficient resources to lead the way on risk assessments. Others stressed the conflicts of interest that would arise, with the regulator unable to both aid water companies in their risk assessment work and act as an independent assessor.

They're not very well resourced. As an organization, we're making the running…It's not very adversarial at all in our case.

They're only interested in when we get it wrong: not when we get it right.

There is an issue of conflict of interest. The Ministry is concerned that whereas its health protection officers used to help water suppliers, it now has to assess what they've done so it can't be helping them at the same time.

The public...has confidence that we know what we're doing. We're pretty well left alone here. The only thing where they might get edgy is that the rate increases are getting a bit high.

There may be individual people who would be interested but I think that might be driven by their particular interest and expertise rather than what their organization is doing. Certainly, I think there is still very much room for individuals to get involved.

Although many in the industry prescribe a high level of stakeholder engagement for better-informed and more defensible decision making, it is clear that the industry has several concerns that need to be alleviated. Finding and communicating effective examples of stakeholder engagement to the industry and stakeholder groups may remedy this situation.

Outsourcing of Risk Expertise

Another viewpoint, voiced strongly by many interviewees, was their unwillingness to outsource risk assessment and management. Most water companies favored the in-house development of expertise and commented on their growing "risk management maturity" that had been gained through direct experience. Consultants were generally used to provide new ideas or for audit purposes:

We do it entirely in-house. Now in the early stages, we did bring in some consultants to sort of help us "get away from dock" so to speak. You have to bring in someone with a briefcase initially, but we're past that stage.

We like to keep it in-house as much as possible...I don't think they've got any particular expertise that isn't available inside...I think when there's big dollars involved, though, people like to get someone external in just to give a second opinion or perhaps a different view of things. That happens fairly regularly.

I wanted to do it in-house because I've seen consultants come, but the intellectual knowledge leaves the company with them...If you're a manager, it [outsourcing] gives you a cop-out. "Oh I don't want to do this workshop today, I'll leave it to the consultant." It's good value to be there.

In one way, we did it on the cheap, but in another way, we did it so we got the knowledge ourselves...We went to a risk consultancy firm, one of the major ones that we'd used previously and we hired two of their top guys for a day, and they took us through a whole range of different risk techniques and processes and explained what we could do with them. We then developed our standard process and our own training package and rolled it out internally ourselves.

We haven't outsourced any of that. It's all done in-house. Auditors [ISO 14001] come through from time to time to look at the risk assessments and hazard assessments.

Barriers to Implementation

Many of the larger utilities cited a variety of cultural barriers as their most significant obstacle to effective implementation. Smaller utilities tended to cite labor shortage as the most significant. One risk manager cited a skill shortage as significant (quotation not included):

I think one of the main barriers is convincing senior managers—they have to buy into it.

There are no cultural barriers. Our City Manager is quite keen on this topic of risk. He's made it a strategic initiative to bring in this capacity.

We are heavily laden with engineers, and engineers are pure seekers of truth, you know?...They don't want...any guidance.

We found it very hard work in doing it...It's really about the culture of the business, changing the way people think, making them start to think about consequences a bit more. At first it is hard work and it does slow things up and you get moaning and groaning about it, but then it becomes almost second nature to them, which is in itself dangerous, of course, as people do start to cut corners off and do tend to take it for granted once they've started using it...We developed our process, "version one," as it was then, and we rolled it out across over a hundred of our assets managers. After a series of three days' training, they all had to select a project they were involved with and apply the technique, then come back and present their findings. This brought out a lot of issues, which we could iron out...it also changed how people thought.

It [risk assessment] is useful, but the problem is that it is a hard-sell. You have folks who see their project as very important and they don't want to hear about it, but once you get past that, it's been useful.

The technology wasn't really a barrier for us because we avoided all technology until we got the methodology right. We made sure we had a fairly robust approach then we built a system to capture it all on.

Sometimes risk gets over-sophisticated...It's a useful tool for us little blokes as long as we can know that there is something in there that...can quickly give us the answers.

Manpower is an issue. You'd always like some more staff...A lot of this [risk management] is aimed at big authorities that have got huge numbers of staff.

It [manpower] hasn't been an issue in risk assessments. It has operationally and has been identified as a risk, but not in this process.

Improving Risk Management Maturity

When asked how a company could improve its risk management strategy, the majority of respondents identified a greater need for the integration of risk management, as previously described. Other areas cited for improvement are included in these excerpts:

We're looking at a number of different things. Two things spring to mind. One thing is to use risk to look for opportunities as well as seeing it as a negative...We're also trying to build a tool which will give us a visual picture of the risk as it cascades through a water supply system from source to tap—to see whether or not the risks do actually pass down the chain or, if they don't at what point does it go away, e.g., what are the risks of delivering a certain volume of water from one stage to the next?

What we need to do, as soon as we can is to get the process to where it's fairly static...We're still adjusting the process as we go, and I have great arguments with my manager...he says, "Settle on a process," and I said, "Look, NASA made mid-course corrections going to the moon: we're going to have to make a few on this."

We've identified a number of things. We're looking at the way we set our thresholds at the moment.

We're trying to document risk management processes at the strategic level and at the operational level. That's the work in progress.

CHAPTER 4
UTILITY SELF-ASSESSMENTS AND BENCHMARKING

Making credible and defensible decisions in organizations requires an institutional capacity to be predictive rather than reactive when managing risk, as well as an aptitude to learn from experience. Implementing risk analysis strategies and decision-making frameworks requires straightforward procedures that can be understood, agreed to, and operated at all levels in an organization.

According to Deloach (2000), the implementation of a working framework should not be seen as the end point of risk management. Businesses must strive to continuously re-assess and improve their capabilities for dealing with risk. Senior management must be able to define their risk management capabilities (both current and desired) and assess the costs and benefits of improvement. Given these ideals, how mature is the water utility sector in assessing and managing risk?

An increasingly applied tool that can assist with this determination is the capability maturity model (CMM). Initially developed for the improvement of software processes (Paulk, Chrissis, and Weber 1993), the model has been adapted and applied for use more broadly (IACCM [International Association for Contract and Commercial Management] 2003), for example, in the oil and gas sector (Sharp et al. 2002). CMMs are practical management tools that codify industry practice so distinctions can be made between organizations, and improvements can then be completed.

WHAT IS A CAPABILITY MATURITY MODEL?

A "capability maturity model" is a tool used to assess the capability and maturity of an organization to perform the key process(es) required to deliver a product or a service. The CMM has its roots in the field of management performance (e.g., Kaplan and Norton 1996) and in quality management systems developed in the 1970s (Crosby 1979, 1996). The strength of the CMM approach is its generic and adaptable framework. This adaptability is leading to increasing numbers of CMMs in other industries and sectors (Fraser, Moultrie, and Gregory 2002; ISO 2000).

A risk management capability maturity model (RM-CMM) for utilities (appendix B) was designed as part of this research. This builds on prior research for design safety capability (Strutt et al. 1998, 1999; Sharp et al. 2002). Similar to earlier models, it is predicated on a belief that measuring and improving the processes involved with managing risk is the most effective manner for reducing risk and improving risk-based decision making. Its focus is on risk management processes (Table 4.1) rather than on outcomes of risk assessment, and it can be implemented at the earliest stages of design, development, or planning. In essence, the RM-CMM provides a strategic risk management framework that

- Enables organizations to establish their current level of maturity for each of the particular risk management processes
- Helps identify the necessary steps to enable the organization to progress to a higher level
- May be used for benchmarking, enabling organizations to compare themselves against other companies in their sector and beyond
- May be used to benchmark the risk management capabilities of key suppliers and partners

Table 4.1
Risk management processes used in the water utility sector

Category	Process	Description
Core business process	Strategic risk planning	The process by which the board, executive management, and senior management identify objectives for corporate risk management and the mechanisms for doing so. The purpose of this process is to guide the risk management activities of the organization and to communicate the corporate risk management strategy to stakeholders.
	Establishing risk acceptance criteria	The process by which a utility sets and assesses the organization's degree of risk significance and prioritizes business risk. The organization is able to rank business risks at periodic risk management committee meetings (or off-line) by reference to these risk acceptability criteria. This process requires an understanding of the relationship between risk and reward.
	Risk analysis	The process by which analytical tasks are performed to identify, assess, and evaluate the risks inherent in business and technical operations.
	Risk-based decision making and review	The process by which the organization selects and develops its responses to risks. The output from this process is the input to risk response.
	Risk response	The process of implementing risk-response strategies, such as the transference, reduction, avoidance, exploitation, and retention of risk.
	Risk monitoring and feedback	The process by which risk data are acquired, compiled, and reported by those responsible. Monitoring activities are largely based on the indicators established to track the progress and effectiveness of risk-response strategies. Monitoring data is transmitted in support of (particularly) risk-based decision making.
	Integrating risk management	The process that involves integration of risk management process interfaces, cross-functional integration of risk management throughout the organization, and integration of risk management with broader business operations.
Supporting processes	Supply chain risk management	The process by which risks from both product supply (e.g., raw materials for manufacturing) and service supply (e.g., outsourcing contracts) are addressed. Managing the risks associated with supplier organizations is critical to maintaining an efficient and reliable service to customers. This is important because of the inherently extended nature of water utility supply chains and the increasing utilization of the outsourcing option.
	Change management	The process by which risk implications of business change programs (e.g., business process reengineering) and technical (e.g., changes in design or technology) change are identified and managed.
Longer-term processes	Education and training in risk management	The process that involves development, maintenance, and enhancement of the skills and knowledge that enable staff to efficiently perform their risk management roles.
	Risk knowledge management	The process that includes gathering and organization of risk information, constant refinement through analysis and collaboration, and dissemination in such a manner as to ensure its effective and consistent use in support of utility decision making.

Case Study 15

Defining Success in Risk Management

An L4 practice in strategic risk planning is the ability to define success in risk management. Two approaches are adopted in this research: the capacity to manage risk and use of performance indicators.

Reflecting on the former, each business unit within one utility performs annual self-assessments of their capacity to manage risk using an externally derived model. Limitations in the feedback obtained from this exercise, however, restrict their maturity to L3, as the interviewee characterized the development of improvement plans to be "dominated by the risk management specialists, rather than being a challenging, innovative dialogue [with the business units]."

Reflecting on the performance indicator approach, it was observed that high-level success was measured in terms of the evolution of corporate risk profiles within the L4 utilities. This was supplemented by discrete performance measures of success in operational risk management (e.g., serviceability in asset management, incident and outbreak measures in drinking water quality management).

Case Study 16

Setting Risk Acceptance Criteria

At L3, risk acceptance criteria are established through a defined process. This process may be characterized as the adaptation of risk ranking techniques. An observed L4 practice is the evolution of risk criteria based on feedback questioning the currency and appropriateness within the utility. For example, one L4 interviewee described how application of their water quality risk ranking technique had led to "risks [being] scored as moderate that had actually hurt the organization in the past." Subsequent adaptation of the criteria focused particularly on "reflecting the knowledge and intuition" of operating staff.

The RM-CMM can be designed and adapted to address all parts and levels in the organization to check the breadth and consistency of knowledge of the different processes and practices throughout an entire organization. Assessments can be implemented either as a self-assessment procedure or by external audit using independent verification authorities. It is believed that the latter, in most cases, gives greater credence to the results of an assessment. Internal assessments, however, are often more useful when using the CMM as an improvement tool rather than a measurement tool. A more detailed description of the RM-CMM is provided by MacGillivray et al. (2006b).

Five levels of capability are defined by reference to the "theory of action" and the concepts of "single- and double-loop learning" (Argyris and Schön 1978). Single-loop learning occurs when risks are detected and the product or service (in this case, the provision of safe water) is amended, allowing the organization to proceed with its objectives. Double-loop learning occurs when risks are managed by modifying the organization's norms, policies, and objectives (case studies 15 and 16).

Table 4.2 shows the interpretations placed on maturity levels (L1 through L5), plus an additional level, L0, which is included because such organizations exist. With these interpretations, organizations can adapt their management processes and organization to optimize delivery.

Table 4.2
Interpretation of maturity levels

Level	Maturity	Mode/style	Process characteristic and effect
5	Optimized	Adaptive—double-loop learning	The organization is "best practice," capable of learning and adapting. It not only uses experience to correct problems but also changes the nature of the way it operates. The processes are continual, explicit components of organizational activity, forming part of the organization's culture.
4	Managed	Quantified—single-loop learning	The organization can control what it does in the way of processes. It determines requirements and ensures that these are met through feedback.
3	Defined	Measured—open-loop learning	The organization can evaluate what it does and how it goes about it but cannot necessarily act on its analyses.
2	Repeatable	Prescriptive	The organization can repeat previous actions but cannot necessarily define those actions.
1	Ad hoc	Reactive	The organization is characterized as a learner organization with complete processes that are not standardized and are largely uncontrolled.
0	Incomplete	Violative	The organization is identified by incomplete processes, resulting in criminal or deliberate violation tendencies.

Double-loop learning is consistent with L5 capability. L4 capability links to single-loop learning in that risk management influences operations. Organizations at L3 and below are "open loop." The L3 organization knows what must be done to deliver acceptable risk but experiences difficulty in using the output of its risk management practice to influence operations. An L2 organization can standardize its procedures and repeat previous actions, but the processes are not properly defined for achieving acceptable risk. An L2 organization may be unclear about which risk management activities need to be performed or they may have limited or inappropriate tools for risk assessment. The lowest-level organizations (L1) are learner organizations; they lack consistency in risk management and may not meet minimum requirements.

WHAT MATURITY LEVEL IS REQUIRED BY UTILITIES?

An obvious question for utilities is what level of maturity is required for effective risk-based decision making? The UKOOA (1999) framework has been amended to illustrate this concept (see Figure 4.1). Many engineering situations require type-A decisions in which risks are managed by adhering to relevant codes or current practices. This can be accomplished by organizations with L2 capability. Most commonly in high-risk operations, however, type-B decisions have to be made. These may involve the use of quantified risk assessment and cost–benefit analysis trade-offs. L3 is suggested as the minimum capability for this category of decision making (case study 17) and ideally L4, particularly in a legislative regime in which companies must demonstrate that risks have been reduced to a level which is as low as reasonably practicable.

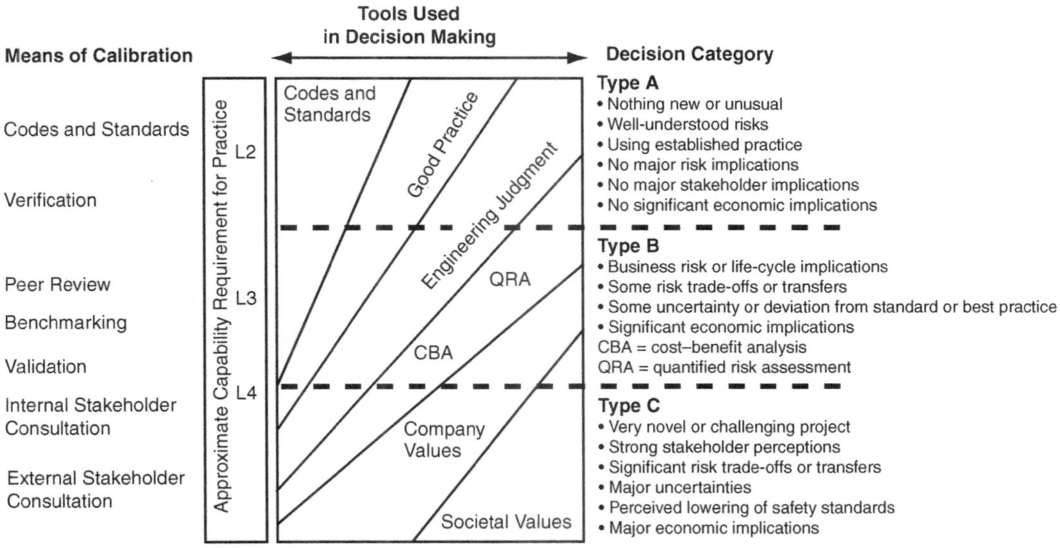

Source: Adapted from UKOOA 1999.

Figure 4.1 Risk-based decision framework showing capability maturity levels required for implementation

Case Study 17

Capability in Managing Risk

An L3 characteristic is the systematic allocation of responsibility, tasks, timescales, guidelines, and resources for implementation of risk-response strategies. These were observed within action plans devised to operationalize risk-based decisions. However, as emphasized by several interviewees, the degree of formality with which these plans are developed is dependent on management style. At L3, lagging indicators are established to verify that implementation proceeds as planned. One interviewee stated that local business unit procedures combined with oversight of the risk register by the risk management team serve this purpose for both asset and business risks. In contrast, another emphasized that the role of a risk manager is not to act as a "central policeman"; hence, responsibility for verifying implementation of decisions is entirely devolved to the business-unit level. In an operational context, a third utility representative provided their "HACCP [hazard analysis and critical control points] verification plan," which details systematic procedures for verifying implementation of their water safety plans.

Where new or potentially controversial activities are proposed or where there are risks with large uncertainties of probability and consequence, type-C decisions predominate. In such situations, it is appropriate to address the concerns, views, and perceptions of all stakeholders, both internal and external. This requires the highest levels of risk management capability and, in particular, an ability to change policies and practices that address type-C decisions. It can be argued, therefore, that organizations with at least an L4 or L5 capability would be needed to respond to these types of risk.

Case Study 18
Supply Chain Risk Management

At one water utility (assessed at L3 in this process), supply chain risk management is explicitly interwoven within procurement policies and procedures. Inspection of their contracting and tendering policy revealed that, in contrast to the much-maligned historical approach of selecting lowest-cost suppliers when basic standards are met, a broad range of prequalification standards are applied as appropriate, including those that address risk explicitly and implicitly. Explicit standards include those addressing commercial risk, health and safety, and delivery risk. Implicit standards include systematic vetting of potential tenders using a range of appraisal methods, including industry-specific and generic-quality standards. Furthermore, supply chain risk is addressed at the selection node. Here, prior to binding acceptance, an evaluation of the probability of failure of the chosen tender to satisfactorily adhere to the contract and the potential effects of such a failure are formally assessed and reported.

From limited data on the supply chain risk management practices of those utilities evaluated at L4, it was observed that although one may require all contractors to utilize a formal risk management framework, another may adopt criteria to ensure that the "risk attitude" of capital partners aligns with their own.

Such organizations have the highest level of assessment capability and double-loop learning potential. They can influence the design and/or adapt their organizational behaviors to meet the challenges raised—for example, in managing risks within their supply chain or through their outsourced activities (case study 18).

CMM DESIGN PROCESS AND METHODOLOGY

The RM-CMM described in this report (appendix B) was developed in close consultation with industry leaders. The lead developer, in concert with a steering group of four expert practitioners, directed the design. Key inputs to the development process were the literature review and interviews, as well as prior knowledge within the steering group of maturity modeling in other utility sectors. The model remains under development. The goal is to render the model fit for industrial use without the need for academic support. This entails balancing the pragmatic needs of the sector and remaining true to the fundamental principles of maturity modeling. Readers may adopt the model to inform an assessment of their in-house risk management capability. A comprehensive version of the RM-CMM is provided in this report (appendix B), although users may prefer to adopt modified versions for their own organizations, particularly when applying the tool within a small water utility where resources may be at a premium.

The benchmarking methodology comprises three components: self-assessment, semistructured interviews, and document analysis. The self-assessment process is by questionnaire and is composed of a series of guideline statements describing how each risk management process is undertaken at the five maturity levels by reference to key attributes (Table 4.3). Responses to the guideline statements are indicated on a Likert-type scale (fully agree, generally agree, partially agree, and disagree), and space is provided for supporting comments.

In total, CMM assessment forms were sent to 16 water utilities in Australia, Canada, the United Kingdom, and the United States. Full responses were obtained from eight utilities; the

Table 4.3
Basic descriptions of the key attributes characterizing risk analysis maturity

Attribute	Attribute description	**Attribute at L3: Risk analysis** *The defined organization—the organization can evaluate what it does and how it goes about it.*	**Attribute at L4: Risk analysis** *The managed organization—the organization can control what it does in the way of processes. It determines requirements and ensures that these are met through feedback.*
Scope	The execution of key practices and the breadth of process implementation throughout the organization	A defined, documented process is in place containing criteria, methods, and guidelines for the identification, assessment, and evaluation (with respect to acceptance criteria) of a broad range of risks throughout core business areas, guided by a risk register. The organization is conversant with and goes beyond the regulatory requirements for risk analysis.	A controlled process is in place containing detailed criteria, methods, and guidelines to manage the identification, assessment, evaluation (with respect to acceptance criteria), establishment of causality, and linking (common cause and dependent) of risks at all levels of the company and across all functional boundaries of the business, guided by a company-specific risk register.
Integration	The level of process embedment within the organization	Procedures are in place to initiate risk analysis processes.	Risk analysis is initiated automatically as part of core business processes (e.g., periodic business risk assessments).
Verification and validation	Verification: ensuring that the process is being performed correctly. Validation: ensuring that the correct process is being performed.	Basic mechanisms are in place to verify that risk analysis is performed as required, largely reliant on lagging indicators. The expertise for validation is generally lacking.	Verification and validation systems are in place to verify the efficiency of risk analysis activities and to validate their expediency (e.g., the organization ensures that the appropriate tools and techniques are being used correctly).
Feedback and organizational learning	The manner in which feedback, both internally and externally sourced, is collected and used to question and revise processes	The risk analysis tool suite is reviewed and modified on an event-driven basis.	Feedback is actively reviewed to improve the execution of risk analysis (e.g., gaps are identified and risk analysis tools and techniques are improved in response to feedback).
Stakeholder engagement	The process of engaging stakeholders, both internal and external, with the purpose of leveraging the process	Risk analysis processes generally reside within the responsible unit, with limited cross-functional or external consultation.	Risk analysis processes generally reside within affected disciplines, and stakeholders work together to define and implement an integrated approach to risk analysis, capitalizing on synergies and collective knowledge.
Competence	The organizational qualities and abilities, both implicit and explicit, that influence process performance	Detailed knowledge of risk analysis resides only within the responsible unit.	Most involved staff exhibit a good level of competence in the selection and application of risk analysis tools and techniques, and have access to support from internal or external expert risk practitioners.

(continued)

Table 4.3 (Continued)

Attribute	Attribute description	**Attribute at L3: Risk analysis** *The defined organization—the organization can evaluate what it does and how it goes about it.*	**Attribute at L4: Risk analysis** *The managed organization—the organization can control what it does in the way of processes. It determines requirements and ensures that these are met through feedback.*
Resources	Extent and use of resources (e.g., people, cost, and tools)	Adequate resources are provided in support of risk analysis, with availability of both qualitative and quantitative tools and techniques.	Sufficient resources are provided in support of risk analysis, a portion of which is made available for risk assessment R&D. A broad range of qualitative and quantitative tools and techniques are available and applied, including methodologies for aggregating and comparing risks.
Documentation and reporting	Documentation and reporting of risk information	Risk analysis outputs are compiled and disseminated in a format that supports decision making.	Risk analysis outputs are compiled and disseminated in a clear, concise, and actionable format that supports real-time decision making, and their reporting is coordinated with other risk-reporting mechanisms (e.g., risk status updates).

maturity levels of an additional four water utilities were then inferred from telephone interviews with their respective risk managers. Additionally, the participation of an electricity utility widely regarded as best practice in risk management was secured with the purpose of identifying transferable good practices.

OVERVIEW OF SAMPLE DATA

Figure 4.2 presents an overview of the self-assessment data supplied by the twelve water utilities assessed:

- Most of the utilities' risk management processes are categorized at L3 or below. Utilities classified their practices as L4 in only a few cases, and most organizations are at or are becoming *defined* (L3) in terms of their risk management processes.
- The profile is relatively mature for the core processes of strategic risk planning and risk analysis (typically L3–L4). However, there appears to be widespread difficulties in transforming risk analysis and monitoring outputs into formats that inform decision makers.
- The profile is relatively immature for the long-term processes (education and training and risk knowledge management; case studies 19 and 20). This reflects the perceived low criticality and immediacy of these processes.

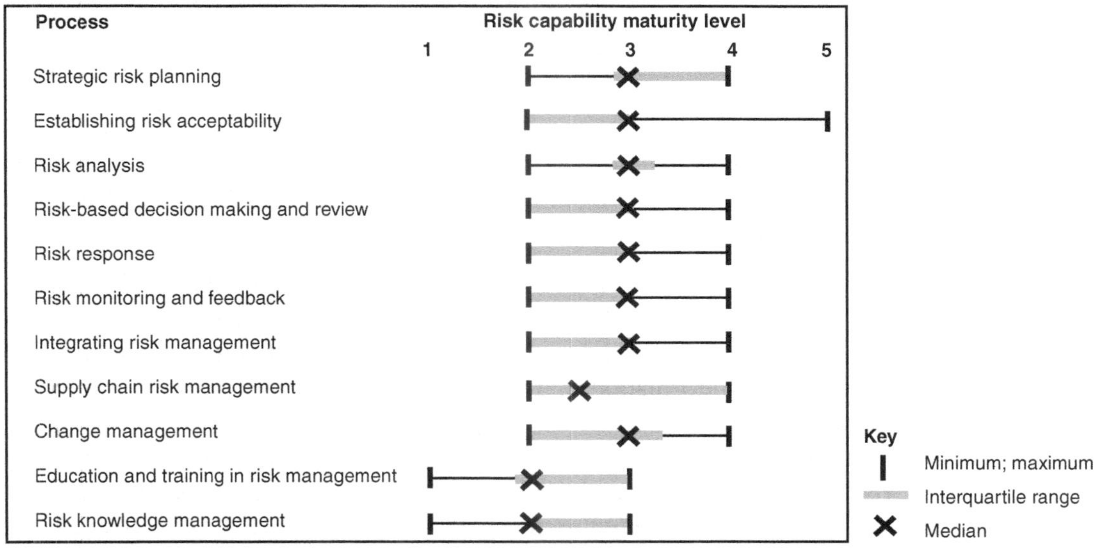

Figure 4.2 Comparison of process maturity profiles of 12 water utilities

Case Study 19

Education and Training in Risk Management

In many L1 and L2 water companies, formalized training and education in risk management is restricted to ad hoc attendance of externally delivered courses and conferences for those who address risk on a daily basis (e.g., seminar attendance by senior managers, HACCP training for operating staff). Greater emphasis is placed on the delivery of on-the-job training, where knowledge transfer occurs through direct participation in the risk management processes (e.g., risk analysis workshops).

One interviewee of a company assessed at L3 in this process described how, in addition to externally delivered HACCP training for key staff, internal training about the fundamentals of drinking water quality management focuses on embedding the risk-based approach in their operating and management principles. Furthermore, postincident analyses highlighted the practical implications of the discipline to operational staff.

In another good-practice example at a different L3 utility, two dedicated risk management training packages exist: an introductory course provided to key strategic members of the business and a more comprehensive program delivered to team leaders. The latter was established following the definition of *risk management* as one of six key competencies required by team leaders. Next, the skill set and knowledge base required for practicing and facilitating effective risk management was defined, which ranges from an understanding of corporate governance to a grasp of the technical aspects of the myriad of analytical techniques adopted. The resultant program, along with the long-standing introductory course, is initiated through formalized procedures, with oversight from local management and the risk team to verify compliance with the established process. Effective delivery of the process is verified through periodic evaluations of the team leaders' abilities to act on the training received. Supplementing these packages, ad hoc training sessions are provided by the risk management team on request. This previously involved conducting workshops with areas of the business that want to improve their risk management processes, holding sessions with key managers, and so forth.

Case Study 20

Managing Risk Knowledge

Risk knowledge management may be subdivided as follows: the storage and access of outputs from risk analysis; risk-based decision making and risk monitoring; and the collection, storage, and access of the raw data that underpins these same processes.

One L2 interviewee stated that data requirements for risk management are not well defined and therefore acknowledged the limited scope of their process: "[We] do not maintain detailed risk information beyond that which is accumulated during risk assessments or that inherent in the normal conduct of business." Furthermore, information sharing is constrained to a problematic extent. This is partly attributable to legislation subjecting all data to public records request, which retards documentation of risk data, and partly due to a "resource-driven inadequacy" relating to how they store, compile, and analyze risk data. In the absence of predefined requirements, risk data collection is typically ad hoc and largely restricted to the requirements of "business as usual." One interviewee described a reliance on expert judgment: "Without senior experienced people, I'm not sure we have the data to underpin [risk management]."

At L3, software packages were used as repositories for the collection, storage, and output data collected throughout the life-cycle of risk management. Currently, most predefined strategies of data collection seem restricted to select operational risks, particularly those whose management is underpinned by formal analytical methodologies (e.g., reliability modeling) or subject to regulatory drivers (e.g., asset management, drinking water quality management). Observed verification practices include the centralized oversight of risk registers and tracking to ensure that predefined data requirements, where established, are met.

OTHER LESSONS FROM BENCHMARKING—SUMMARY

Results from the benchmarking trial are further analyzed by MacGillivray et al. (2006b, c) and Strutt et al. (2005). The following general observations can be made.

- The range of methodologies used for risk analysis by water companies is limited, suggesting a need for training in assessment methods and competency improvements.
- Verification of risk management processes is weak and validation activities are largely absent. There seems to be a wait-and-see approach to validation of risk assessments.
- Although company procedures to initiate risk management processes exist, integration within core business processes remains elusive. In all areas, there is a limited understanding of how the risk management processes interrelate.
- Engagement of stakeholders in the process, both internal (e.g., cross-functional groups) and external (e.g., regulators), is limited.
- The inability to measure the value of risk management as a process has restricted buy-in.
- In many cases, risk tolerance is weakly defined, often being cited as a board-level issue. Lack of risk acceptance criteria can adversely impact risk-based decision making and risk response.

- In some cases, emphasis on individual units and departments "owning" the processes has led to a proliferation of disparate risk initiatives. The processes are not coordinated in a strategic manner. This problem can also impose a barrier to integrated risk management.
- Although executive management generally buys into the central role of risk management, communicating its status throughout the company proves somewhat difficult. "Getting it [risk management] part of the mindset" is often cited as the core function of the risk team.
- As organizations mature, the widespread absence of "learning cultures" will prove to be the key constraint in proceeding beyond L4. This seems to be largely beyond the control of corporate risk management teams (RMTs).

CHAPTER 5
RISK MANAGEMENT IMPROVEMENT PLANS

In order to test the potential of the RM-CMM as an improvement tool, case studies were conducted at two utilities participating in the project. Figure 5.1 displays the two risk management maturity profiles. Both organizations are relatively mature for the core and supporting processes of risk management, attaining L3 (defined) in most of these processes.

UTILITY 1

Although utility 1 has succeeded in implementing sophisticated tools for risk management, its profile reflects a self-acknowledged transition status toward a fully mature ERM organization.

Strategic Risk Planning

The focus of strategic risk planning was on measuring the performance of risk management within the utility, with the belief that placement of risk metrics within a structured model might improve uptake of the company risk framework. Two key stages were identified: definition of criteria for evaluating the performance in risk management (e.g., variability in key risk exposures) and the identification of metrics to support this evaluation (metrics relating to the evolution of risk exposures), as presented by MacGillivray et al. (2006c).

Establishing Risk Acceptance Criteria

The principal limitation identified in establishing risk acceptance criteria was the absence of an overarching policy for the determination of risk acceptability. After stakeholder consultations, the cost–benefit principle of risk reduction was agreed upon and the RMT proceeded to communicate this throughout the business, along with exceptions to the rule (e.g., adoption of the ALARP [as low as reasonably practicable] criteria relating to health and safety issues). Building on an existing asset management program, a methodology was developed to derive the probable cost of risk based on a series of single-point estimates.

Risk Analysis

The range of tools available for risk analyses was not documented, nor were criteria for tool application uniformly available. The RMT was tasked with identifying and documenting the instruments in the tool kit, defining criteria for their application, and establishing formal procedures for their use. Mechanisms for verification that risk analysis was performed as required were to be adapted and strengthened. A formal gap analysis and improvement of the risk analysis tool kit was advised. A preliminary recommendation was to devolve the generic strategic scoring system to arrange systems at the business unit level. Finally, the development of stakeholder maps detailing the potential contributions and requirements of key internal and external factors was undertaken.

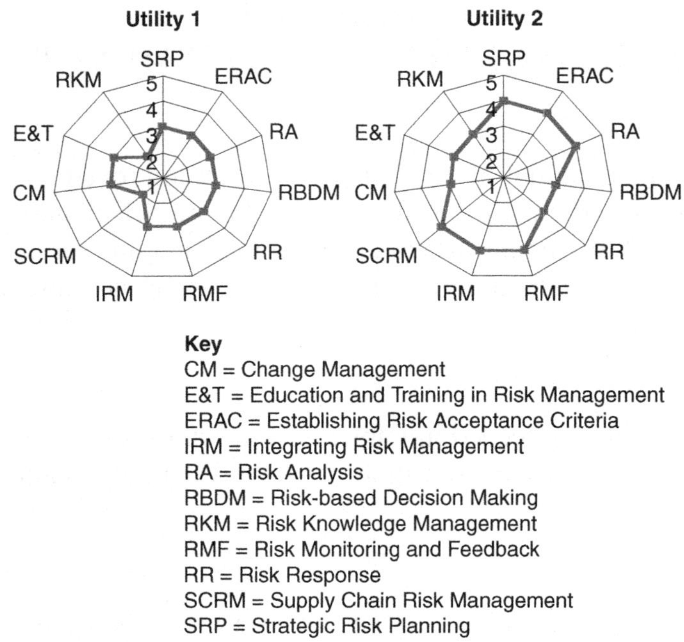

Figure 5.1 Maturity profiles of two water utilities

Risk Response

In the risk-response process, recommendation was passed to the RMT to define the range of risk-response strategies (e.g., to treat, terminate, tolerate, transfer, and exploit) and to develop templates providing examples of how these strategies may be applied. Recommendations were made to develop review mechanisms for strengthening validation at the decision-making and implementation stages. The intention was to filter out the tendency to treat all risks, to introduce greater rigor and defensibility within decision making, and to improve the capacity to learn organizational lessons.

Education and Training in Risk Management

Education and training in risk management was of particular interest to the RMT, who viewed it as both a means of improving staff competencies and awareness with subsequent influence on cultural change. The first recommendation was to identify the education and training requirements of staff in managing risk, considering both technical (e.g., basic training in the application of risk analysis tools) and broader managerial (e.g., basic understanding of risk management concepts, policies, and procedures) competencies. A gap analysis of existing education-and-training programs was suggested. Therefore, a remedial plan was to be produced, detailing the education-and-training vehicles to be adopted (e.g., classroom training, workshops, structured on-the-job training), the competencies they are intended to impart, staff responsibilities, and time scales.

Risk Knowledge Management

The focus of risk knowledge management was on determining and capturing the data sources that improve risk management. The initial requirement was to define the key risk categories from the strategic to operational levels of the organization, followed by a mapping of potential information sources (e.g., databases, initiatives, projects) linked to those categories, and finally to integrate these information sources with risk analysis, monitoring, and decision-making processes (e.g., through raising employee awareness, reengineering information management processes).

UTILITY 2

In the case study for utility 2, the focus was on the role of risk management within various business functions of the utility. The application of risk management strategies within the process engineering, water quality management, and occupational health and safety functions are discussed.

Risk Analysis

Process Engineering

The identification and evaluation of the potential for deviations from utility design specifications were examined. Formal application of risk analysis in this utility is restricted to dam and process engineering. Beyond these exceptions, logically, the systems are sufficiently familiar and well characterized that adherence to codes and standards is deemed sufficient. However, where systems are characterized by greater levels of uncertainty with a potential to deviate from routine operation, particularly where there are public health implications, recourse to formal risk analysis is required.

In process engineering, risk analysis was observed in the application of hazard and operability (HAZOP) studies. HAZOP studies systematically evaluate the process and engineering intentions of new or existing facilities in order to identify the hazards that may arise because of deviations from design specifications. Typically, a carefully selected team examines a process (e.g., disinfection) subdivided into nodes. At each node, the team applies guide words (e.g., low) to process parameters (e.g., ozone levels) to identify ways in which the process may deviate from its design intention; this occurs before evaluating the causes and consequences of the deviation.

Risk analysis in process engineering was observed to be in transition from repeatable to defined practice because the utility was in the process of formalizing technical guidelines for the application of HAZOP studies. Before this transition, there had been a reliance on the judgment of experienced engineers and project managers to initiate and facilitate the process. In other words, staff had applied HAZOP studies before and could be relied on to replicate past experience. Defining the process became more urgent so that an infrastructure could be created which supports its execution beyond the tenure of experienced staff. This is particularly important in an industry characterized by high levels of restructuring and staff turnover.

In practice, this formalization entails specifying the initiation criteria for a HAZOP study (e.g., when and in what circumstances it should be applied), formalizing the procedures for its execution (e.g., detailing the roles, responsibilities, tasks, and activities required for effective undertaking), and providing staff with the requisite expertise (e.g., workshops or external training). With these elements in place, the process techniques can be applied even though experienced staff may have left the program.

A further improvement would be to implement process control (verification and validation). Verification involves ensuring procedural compliance (e.g., supervisory sign-off) and providing quality assurance of its execution. In this context, quality assurance may involve technical peer reviews. The following types of questions may be included in such a review:

- Did the analysts systematically work their way through the HAZOP study or did they overlook important scenarios, components, and process flows?
- Were all stages and operating modes of the process considered (e.g., startup, shutdown, and transitioning to partial operation)?
- Was adequate time spent on the analysis?

Validation seeks to address whether the process itself is correct, regardless of its execution. For example,

- Does application of a HAZOP study constitute best practice in process engineering or should it be supplemented by quantitative techniques?
- Is it appropriate to rely on codes and standards within other engineering disciplines or should formalized application of risk analysis be extended?

Avenues for addressing these questions practically are benchmarking and external auditing. Both seek to examine current practices with respect to industry leaders. There is a trade-off between the investments required to become the leading edge in risk analysis and the benefits to be obtained. Risk analysis remains an expert discipline, and many organizations are more comfortable with historic and proven implicit approaches.

Drinking Water Quality Management

Utility 2 applies an adapted form of the hazard analysis and critical control points (HACCP) methodology. The process begins with the division of the water system into discrete subsystems (e.g., catchment, pretreatment, distribution). Within each subsystem (e.g., catchment) the sources of risk to water quality (e.g., native animals) are identified and assessed for their maximum risk (i.e., without considering the barriers in place). An equation is applied to determine the residual risk (i.e., having considered the barriers in place). This equation is weighted according to the inherent capacity of each barrier (e.g., ultraviolet disinfection) to reduce the hazard loading and how well each barrier is designed, operated, and managed.

Utility 2 was concerned with defining the process. This entailed specifying initiation criteria for applying the methodology (e.g., cyclical criteria, and reactive initiation points as the environment or barrier systems change), formalizing the procedures for its execution (e.g., detailing the roles, responsibilities, tasks, and activities required for its effective undertaking), and providing staff with the requisite expertise (e.g., workshops or external training). A valuable addition would be to specify the information inputs to the assessment (e.g., at treatment: process flow and instrumentation diagrams; at catchment: geological and land use maps, and raw water quality parameters). Predefining the data inputs removes the human tendency to rely on expert judgment and underpins the qualitative assessment with objective data analysis. In large part, this is the codification of past practice in applying the methodology. Codification is particularly important as responsibility for undertaking the process is set to be devolved to regional and operating staff. Similarly, process control is desirable, and the principles discussed in the "Process Engineering" section apply here as well.

Risk-based Decision Making

Occupational Health and Safety

The identification and evaluation of solutions to manage occupational health and safety risks are of concern in risk-based decision making. The most important point here is that risk analyses are structured in such a way as to explicitly examine the underlying mechanisms through which health and safety hazards may be realized, thus informing decisions on their practical management.

The identification of solutions is further informed by a hierarchy of health and safety controls: engineering (e.g., substitution, isolation, modification to design, guarding, and mechanical ventilation), administrative (e.g., safe work practices, procedures, training, supervision, and nominating maximum exposure times), and protective personal equipment. Once identified, solutions are evaluated with reference to criteria that include cost, feasibility, and risk reduction achieved.

What then is the value of this structure? In its absence, decision makers would be bereft of practical guidance for identifying and evaluating risk reduction options and would likely find themselves relying on past experiences to generate solutions. With such an approach, health and safety would be under pressure when circumstances change, whether through the introduction of new technical processes or modifications to work practices.

Process control was further observed in the form of peer reviews at decision-making points (e.g., engaging occupational health and safety practitioners in novel situations) and post-implementation audits that sought to evaluate the operational impact of risk reduction strategies.

CHAPTER 6
SYNTHESIS AND ANALYSIS

RISK ANALYSIS STRATEGIES

From embedding corporate governance to the management of individual assets, the ability to understand, communicate, assess, and manage risk has become a mainstream business activity. Many utilities now integrate their responsibilities for financial control alongside their risk management programs, including those for asset management and regulatory compliance. As these strategies for risk analysis develop and intertwine, keeping the core business objectives close at hand is critical if one is to manage the risks with the greatest consequences.

Public health protection is the principal goal of a drinking water utility, delivered through supplying safe water. Clearly, the risk factors that might compromise this overarching goal can lie undetected within organizations, acting as precursors for potential water quality incidents. Managing risk, therefore, is more complex than optimizing asset replacement, ensuring compliance, or managing chemicals in catchments. It requires vigilance at all levels within organizations, a focus on unambiguous business goals, and thus an organizational commitment toward attaining higher levels (L3, L4) of risk management maturity.

The modern operating environment for water utilities is one of considerable regulatory and market scrutiny. Alongside managing the risks to their businesses, utility managers must be able to take risks that allow technological innovations to proceed and new organizational structures to develop. For many utilities, the goal of public health protection may come under pressure from a plethora of other business risks. Trade-offs are a necessary requirement of business risk management, but the clarity of corporate objectives may suffer under the complexities of regulatory incentives, rapidly changing organizational structures and the split accountabilities caused by outsourcing, for example.

RISK CULTURE AND MINDFULNESS

One theme that emerges from this research is the necessity to progress from risk analysis and risk management frameworks to embedding of a risk management culture within organizations. The structured interviews completed as part of this research provided candid insights into how risk frameworks and decision process schematics are viewed on the ground in water utilities. Even more challenging than securing recognition for these systems is the generation of a risk management culture within an organization.

Making risk management permanent requires the changing of hearts and minds so that responsibilities are shared, corporate knowledge is managed effectively, and essential communication channels are opened in the event of an incident. Much has been written on safety culture in the light of organization accidents since the 1970s, much of it coincident with good risk management practice. Developing a risk management culture that is sustaining and continues to learn and improve in the face of the inevitable peaks and troughs of organization performance requires the following components (Taylor 2005):

- Leadership, formal procedures, and an appetite for conservative decision making where safety comes first, even under pressure

- A culture of sharing reported close calls
- Good communication at the appropriate level
- An open, learning, organizational culture able to benchmark itself against the best-in-class, systematic competency checking
- Effective management of organizational change and the ability to prioritize

Disasters and incidents have deeply rooted causes (Reason 1997, Hurst 1998) that are typically a combination of technical failures and an incapacity to manage change. Cataloged for waterborne disease outbreaks in the drinking water sector by Hrudey and Hrudey (2004), such events result from a failure to convert hindsight into foresight and typically occur during the following types of situations (Taylor 2005):

- When there is a loss of institutional foresight and corporate memory
- In the face of strong market pressure for efficiency gains
- When there are considerable elements of outsourcing
- Where organizations fail to maintain their status as an "intelligent customer"
- With loss of internal technical expertise
- Particularly during or after periods of business reengineering

Accidents can be generated because of cost pressures, priority-based work ethics, and rushed changes. For organizations to become resilient and mindful, they must be able to anticipate and circumvent threats to corporate objectives and manage severe pressures and conflicts between performance and the risks that threaten it. Modern management culture sets a strong impetus on doing more for less (the "lean" model) and on maintaining business continuity. Middle managers may find it difficult to challenge this philosophy, yet when they do, risks that are not easily quantified in monetary terms may be allocated restricted time at board meetings and then lie dormant within the organization as latent causal factors (Reason 1997). Leadership is key to establishing the right culture in terms of the expectations and examples that are set.

CHAPTER 7
RECOMMENDATIONS FOR WATER UTILITIES

This research has raised a series of practical questions about how organizations, and water utilities in particular, endeavor to embed risk management within their organizational structures. How do organizations begin to develop a risk management culture without first having to suffer a major accident? How do organizations force themselves to ensure that risk issues are treated seriously? How can the volumes of risk data, information, and knowledge gathered by risk managers be usefully processed in light of accident/incident prevention? When should executive-level managers listen to challenges from employees? Above what threshold should they act? Arguably, are risk managers an additional source of risk, because in taking institutional responsibility for coordinating risk assessment and management, they absolve others of their individual responsibilities for risk management?

Managing risk requires wisdom and reflection, and preventive approaches that are creative and forward-looking. The best-in-class organizations are mindful of risks to their operations. Weick and Sutcliffe (2001) characterize this mindfulness in organizations that demonstrate the following characteristics:

- Preoccupation with failure and its root causes
- Reluctance to (over)simplify
- Sensitivity to operations
- Commitment to resilience
- Deference to expertise

Securing mindfulness for the water utility sector is critical to the implementation of risk management. It is vital that water companies proactively seek out new hazards, not just offer analysis to support current practices. The following attributes are proposed for water utilities seeking to develop mindfulness (Hrudey, Hrudey, and Pollard 2006):

- Informed vigilance is actively promoted and rewarded.
- An understanding of the entire system exists, and knowledge of its challenges and limitations are promoted and actively maintained.
- Effective, real-time treatment process control, based on understanding critical capabilities and limitations of the technology, is the basic operating approach.
- Fail-safe multibarriers are actively identified and maintained at a level appropriate to the challenges facing the system.
- Close calls are documented and used to train staff about how the system responds under stress and to identify what measures are needed to make such close calls less likely in the future.
- Operators, supervisors, lab personnel, and management understand that they are entrusted with protecting the public's health and are committed to honoring such responsibility above all else.
- Operational personnel are afforded the status, training, and remuneration commensurate with their responsibilities as guardians of the public's health.

- Response capability and communication are improved, particularly as post-9/11 bioterrorism concerns are addressed.
- An overall mentality of continuous improvement and total quality management pervades the organization.

CHAPTER 8
CONCLUSIONS

Although the water utility sector has developed risk analysis strategies, it is struggling to use risk analysis data within the broader context of utility decision making. Widespread difficulty in ensuring that risk analysis informs decision making is apparent. The following conclusions have resulted from this research:

- A tendency remains to view risk assessment as an end rather than a means to inform company exposures and opportunities for innovation.
- Risk management requires greater advocacy in the sector and within individual organizations.
- The promotion of water safety plans is driving an integrated approach to public health risk analysis from catchment (watershed) to tap. Implementation of the water safety plan approach is becoming a powerful vehicle for a renewed emphasis on risk management within water utilities.
- Risks and values are intimately linked; however, there is little explicit expression, at present, of the risks that utilities are prepared to accept, whether the utilities are private, public, or corporatized government structures.
- Failure to manage risk in a climate of efficiency gains may leave the sector exposed to the types of organizational disaster witnessed during a fatal outbreak of waterborne illness in Walkerton, Ont. Regardless of its complexity, the tolerance of risk within an organization must be discussed and confronted.
- Securing a risk management culture on the ground within a water utility requires effective risk knowledge management, tenacity in addressing the underlying causes of incidents, and strong leadership. Further research that addresses risk management cultures within the sector is warranted.
- The compelling case for effective risk management should be the core argument of water utilities for moving toward an organizational structure that can be more strategic and forward-thinking.

APPENDIX A
STRUCTURED INTERVIEW QUESTIONNAIRE

DECISION-MAKING PROCESS

What sort of company structures and accountabilities do you have in place for risk management?

Have you or anyone else in the company had any formal training in risk assessment or management?

Does your company have any formal documentation on risk assessment?

Do you use any formal risk frameworks in your company?

For risk assessment/risk management/quality management? What about water quality management plans e.g., NHMRC or WHO water safety plans?

What has been your experience of these to date?

How do you allow for values and opinions in the decision-making process?

What about where there are conflicts between different management priorities?

(e.g., water quality vs. water conservation issues)

Can you provide examples of how risks are assessed and managed across strategic, programme and operational interfaces?

(May have answered part of this already).

STAKEHOLDER INVOLVEMENT

Do stakeholders have any participation in your risk assessments?

Do you think there is much of an appetite amongst stakeholders to get involved in the risk assessment process?

Have you encountered any differences in the perception of risk taken by the public and business?

(e.g., with regard to the storage of chlorine gas, for example)

RISK TOOLS AND TECHNIQUES

Can you tell me anything of the journey you've travelled on with regard to risk management?

(e.g., how did your company first become interested in risk?)

Do you have any good or bad examples of risk assessment use that you'd be prepared to tell me about (good or bad)?

Could you tell me a bit about any other risk tools you're currently using or developing?

What successes and failures have you had implementing these tools?

From your experience what do you think are the main barriers to using these within your organisation?

Culture barriers?
Technological barriers?
Skills shortage?
Manpower shortage?

Is most of your risk assessment work conducted in-house or do you outsource expertise?

Why outsource?

How do you think your company could do its risk assessments better?

OTHER

Is there anyone else within the company I could contact and talk to, regarding specific issues?

Anything else you want to add?

APPENDIX B
RISK MANAGEMENT CAPABILITY MATURITY MODEL (VERSION 0.1)

Process (Description)	Maturity level	Five maturity levels from learner to best practice*	Fully agree	Generally agree	Partially agree	Disagree	Evidence (please use separate sheet if necessary)
Strategic risk planning This is the process by which the board, executive management, and senior management identify their objectives for corporate risk management and the mechanisms for doing so. The purpose of this process is to guide the organization's risk management activities and to communicate the corporate risk management strategy to stakeholders. Key tasks include 1. Developing a corporate vision for risk management. 2. Establishing a risk management framework setting out the core roles, responsibilities, accountabilities, and mechanisms for risk management throughout the organization. 3. Identifying key risk and opportunity elements at the strategic level. 4. Developing corporate strategy to manage identified risk and opportunity elements. 5. Allocating resources for implementing a corporate risk management strategy. 6. Establishing risk policies and guidelines to inform operational procedures. 7. Developing criteria indicating how success and failure in risk management will be measured at the corporate level.	LEVEL 5–Optimized The organization is "best practice," capable of learning and adapting. It not only uses experience to correct problems but also changes the nature of the way it operates.	**Scope:** Strategic risk planning and the associated risk framework (detailing core roles, responsibilities, accountabilities, and mechanisms for risk management) address all types of risks throughout all areas of the organization, with procedures in place for the management of opportunities as well as adverse consequences. Risk policies and guidelines enable the alignment of operational procedures with long-term risk strategy.					
		Integration: Strategic risk planning is fully aligned with broader business planning, with explicit links established between risk management objectives and corporate objectives (e.g., supply and demand objectives incorporate measures of uncertainty).					
		V+V: Management continually establishes measurable targets for improving risk management, with systems in place to verify their achievement and validate the expediency of their enabling strategies.					
		FB/OL: Feedback is actively used to improve risk planning and the development of strategies and policies, enabling changes to organizational structures and practices to optimize conditions for implementing the risk management strategy.					
		FB/OL: Risk management goals, objectives, and related plans are proactively modified in light of internal and external changes and influences on the business.					
		Competence: The organization is conversant with international practice in strategic risk planning and is active in improving this, with all involved staff, including executive management, fully cognizant of the methods and implications of strategic risk planning.					
		Resources: Resources for implementing risk strategy are sufficient and are optimally and flexibly allocated in response to need, and the organization applies formal cost–benefit analysis in support of resource allocation (i.e., the "value" of different strategies can be quantified, then their implementation can be accordingly resourced).					
	LEVEL 4–Managed The organization can control what it does in the way of processes. It determines requirements and ensures that these are met through feedback.	**Scope:** Strategic risk planning and the associated risk framework (detailing core roles, responsibilities, accountabilities, and mechanisms for risk management) address risks at all levels of the company and across all functional boundaries of the business. Established policies and guidelines direct the implementation of risk strategy at the operational level.					
		Integration: Strategic risk planning is automatically initiated as part of routine business planning and is coordinated with broader business planning (e.g., risk planning and broader business planning cycles are aligned).					
		V+V: Risk management goals and objectives have measurable performance indicators, with systems in place to track progress in their achievement and validate the expediency of their enabling strategies.					
		FB/OL: Senior management reviews the validity of risk strategies and policies on a periodic and event-driven basis, making modifications based on feedback from the risk management committee, corporate experiences, regulatory input, etc.					
		SE: Cross-functional teams and key external stakeholders work together to define and implement an integrated approach to risk planning, and core plans and strategies are endorsed within individual business units by key external stakeholders (e.g., shareholders, regulators).					
		Competence: The organization is conversant with sectoral practice in strategic risk planning and makes efforts to exceed this, with most involved staff aware of the methods and implications of strategic risk planning.					
		Resources: Sufficient resources are provided to implement a risk management strategy.					
	LEVEL 3–Defined The organization can evaluate what it does and how it goes about it.	**Scope:** Strategic risk planning is defined and documented, resulting in a tailored risk framework (detailing core roles, responsibilities, accountabilities, and mechanisms for risk management), addressing a broad range of risks that is applied throughout core business areas. Policies and guidelines are established for risk management.					
		Integration: Procedures are in place to initiate strategic risk planning.					
		V+V: Basic mechanisms are in place to verify progress in the achievement of risk management goals and objectives, largely reliant on lagging indicators. The expertise for validation is generally lacking.					
		FB/OL: Risk strategies and policies are reviewed on an event-driven basis and streamlined.					
		SE: Strategic risk planning responsibilities generally reside within one group (e.g., executive management), with limited cross-functional or external consultation.					
		Competence: The organization is conversant with regulatory requirements for risk management and their planning exceeds these boundaries, with detailed knowledge of the methods and implications of strategic risk planning resident with senior-level (i.e., experienced) responsible staff.					
		Resources: Adequate resources are provided to implement risk management strategy.					
	LEVEL 2–Repeatable The organization can repeat previous actions but cannot necessarily define those actions.	**Scope:** Basic process established for strategic risk planning, focused on meeting basic regulatory requirements and addressing "mission critical" risks.					
		Integration: Strategic risk planning is reactive, often initiated in response to an event or situation.					
		FB/OL: Strategic risk planning is based on regulatory guidance and previous experience, regardless of applicability or currency.					
	LEVEL 1–Initial	There is no formal process in place for strategic risk planning. An ad hoc approach exists. There is limited awareness of regulatory requirements for risk management.					

Process (Description)	Maturity level	Five maturity levels from learner to best practice*	Fully agree	Generally agree	Partially agree	Disagree	Evidence (Please use separate sheet if necessary)
Establishing risk acceptance criteria Establishing risk acceptance criteria is the process by which a utility assesses the organization's degree of risk significance and prioritizes business risk. The organization is able to rank business risks at periodic risk management committee meetings (or off-line) by reference to these risk acceptability criteria. This process requires an understanding of the relationship between risk and reward. Key tasks include 1. Defining high-level risk areas for the business and assets. 2. Defining the level of risk that is acceptable to the various stakeholders among key risk categories (e.g., financial risk thresholds, technical requirements for safety and reliability, ALARP, etc.). 3. Allocating risk acceptance criteria to key components of the business and assets.	LEVEL 5–Optimized The organization is "best practice," capable of learning and adapting. It not only uses experience to correct problems but also changes the nature of the way it operates.	**Scope:** Risk acceptance criteria are developed for all types of risk throughout all areas of the organization, and their establishment is often informed by formal cost–benefit analysis (e.g., for key assets).					
		Scope: Criteria are in place to guide the balancing of business opportunities with adverse consequences (e.g., detailing whether international expansion is acceptable in risk terms).					
		Integration: The development of risk acceptance criteria is a continual, explicit component of all business processes, forming part of the organizational culture.					
		FB/OL: Feedback provides the organization with an understanding of the relationship between its established levels of risk acceptance and business performance (i.e., the relationship between risk and reward). Risk acceptance criteria evolve based on this knowledge to provide a more effective balance of risk and reward.					
		FB/OL: Risk acceptance criteria are proactively modified with regard to internal and external changes and influences on the business.					
		Competence: The organization is active in developing international risk acceptance criteria to form best practice, with all involved staff, including management, fully cognizant of the methods and implications of developing risk acceptance criteria.					
	LEVEL 4–Managed The organization can control what it does in the way of processes. It determines requirements and ensures that these are met through feedback.	**Scope:** Risk acceptance criteria are developed, with an appropriate degree of quantification (e.g., for key assets) and consistency, for risks at all levels of the company and across all functional boundaries of the business.					
		Scope: Risk acceptance criteria are allocated at the asset and business unit levels.					
		Integration: Risk acceptance criteria are automatically developed as part of core business processes.					
		V+V: Systems are in place to verify that risk acceptance criteria are uniformly adhered to and to validate the expediency of established criteria.					
		FB/OL: Feedback is actively used to revise the process through which risk acceptance criteria are developed.					
		SE: Cross-functional staff and key external stakeholders (e.g., regulators) provide input to the process of developing risk acceptance criteria.					
		Competence: The organization is conversant with sectoral practice in developing risk acceptance criteria and makes efforts to exceed this, with most involved staff aware of the methods and implications of developing risk acceptance criteria.					
	LEVEL 3–Defined The organization can evaluate what it does and how it goes about it.	**Scope:** A defined, documented process is in place to develop risk acceptance criteria for a broad range of risks throughout core business areas.					
		Scope: Risk acceptance criteria are allocated at the asset level.					
		Integration: Procedures are in place to initiate the development of risk acceptance criteria.					
		V+V: Basic mechanisms are in place to verify adherence to risk acceptance criteria, largely reliant on lagging indicators. The expertise for validation is generally lacking.					
		FB/OL: The organization reviews and modifies risk acceptance criteria on an event-driven basis.					
		SE: The responsibility for developing risk acceptance criteria generally resides within one group (e.g., risk management committee or executive management), with limited cross-functional or external consultation.					
		Competence: The organization is conversant with regulatory requirements for risk acceptance (e.g., risk-based water quality standards) and makes efforts to exceed these, with detailed knowledge of the methods and implications of developing risk acceptance criteria resident with senior-level (i.e., experienced) responsible staff.					
	LEVEL 2–Repeatable The organization can repeat previous actions but cannot necessarily define those actions.	**Scope:** A basic process is in place to develop risk acceptance criteria, generally through regulation and standards. The organization may define its own acceptance criteria for "mission critical" risks.					
		FB/OL: Risk acceptance criteria are retrospective and historical, regardless of applicability or currency, with limited organizational ability to plan ahead or learn from experience.					
	LEVEL 1–Initial	There is no formal process in place to develop risk acceptance criteria. An ad hoc approach exists. There is an inappropriate use of standards.					

Process (Description)	Maturity level	Five maturity levels from learner to best practice*	Fully agree	Generally agree	Partially agree	Disagree	Evidence (Please use separate sheet if necessary)
Risk analysis In this process analytical tasks are performed to identify, assess, and evaluate the risks inherent in business and technical operations. Key tasks include 1. Identifying risks inherent in business and technical operations. A range of techniques are available for this purpose (e.g., what-if analysis, scenario planning, hazard and operability studies, hazard analysis and critical control points, etc.). This task should be informed by a risk register that is continually updated. 2. Assessing risks for likelihood of occurrence, likelihood of consequences, and overall impact on the utility's corporate objectives. A range of techniques are available for this purpose (e.g., failure mode and effects analysis, failure mode and effects criticality analysis, net-present-value-based analyses, compliance assessments, pollutant modeling tools, etc.) 3. Evaluating assessed risks with respect to defined acceptance criteria.	LEVEL 5–Optimized The organization is "best practice," capable of learning and adapting. It not only uses experience to correct problems but also changes the nature of the way it operates.	**Scope:** The organization is regarded as best practice in risk analysis, both nationally and internationally, with comprehensive and proactive identification, assessment, evaluation (with respect to acceptance criteria), establishment of causality, and linking (common cause and dependent) all types of risk throughout all areas of the organization, using a "fit-for-purpose" range of tools and techniques.					
		Scope: Risk analysis extends to include the risk-based evaluation of business opportunities.					
		Integration: Risk analysis is a continual, explicit component of all business processes, forming part of the organizational culture.					
		V+V: Management continually establishes measurable targets for improving the risk analysis process, with systems in place to verify their achievement and validate the means through which they are pursued.					
		FB/OL: Feedback is actively used to improve both the execution and philosophy of risk analysis, and further enables the adaptation of organizational structures and practices to optimize its ability to perform risk analysis.					
		Competence: High level of competence among all staff involved in risk analysis, including management, with extensive experience/expert advice used to assist the selection and application of the most appropriate tools and techniques for risk analysis (e.g., the organization knows where and when to source expert assistance, worldwide).					
		Resources: Resources for risk analysis are sufficient and are optimally and flexibly allocated in response to need.					
		Resources: The organization is active in researching and developing improved tools and techniques for risk analysis, with appropriate resources made available.					
	LEVEL 4–Managed The organization can control what it does in the way of processes. It determines requirements and ensures that these are met through feedback.	**Scope:** A controlled process is in place containing detailed criteria, methods, and guidelines to manage the identification, assessment, evaluation (with respect to acceptance criteria), establishment of causality, and linking (common cause and dependent) of risks at all levels of the company and across all functional boundaries of the business, guided by a company-specific risk register.					
		Integration: Risk analysis is automatically initiated as part of core business processes (e.g., periodic business risk assessments).					
		V+V: Verification and validation systems are in place to verify the efficiency of risk analysis activities and validate their expediency (e.g., the organization ensures that the appropriate tools and techniques are being correctly used).					
		FB/OL: Feedback is actively reviewed to improve the execution of risk analysis (e.g., gaps are identified and risk analysis tools and techniques are improved in response to feedback).					
		SE: Risk analysis processes generally reside within affected disciplines, and stakeholders work together to define and implement an integrated approach to risk analysis, capitalizing on synergies and collective knowledge.					
		Competence: Most involved staff exhibit a good level of competence in the selection and application of risk analysis tools and techniques, and have access to support from internal or external expert risk practitioners.					
		Resources: Sufficient resources are provided in support of risk analysis, a portion of which is made available for risk assessment R&D. A broad range of qualitative and quantitative tools and techniques are available and applied, including methodologies for aggregating and comparing risks.					
		D+R: Risk analysis outputs are compiled and disseminated in a clear, concise, and actionable format that supports real-time decision making, and their reporting is coordinated with other risk-reporting mechanisms (e.g., risk status updates).					
	LEVEL 3–Defined The organization can evaluate what it does and how it goes about it.	**Scope:** A defined, documented process is in place containing criteria, methods, and guidelines for the identification, assessment, and evaluation (with respect to acceptance criteria) of a broad range of risks throughout core business areas, guided by a risk register. The organization is conversant with and goes beyond the regulatory requirements for risk analysis.					
		Integration: Procedures are in place to initiate risk analysis processes.					
		V+V: Basic mechanisms are in place to verify that risk analysis is performed as required, largely reliant on lagging indicators. The expertise for validation is generally lacking.					
		FB/OL: The risk analysis tool suite is reviewed and modified on an event-driven basis.					
		SE: Risk analysis processes generally reside within the responsible unit, with limited cross-functional or external consultation.					
		Competence: Detailed knowledge of risk analysis resides only within the responsible unit.					
		Resources: Adequate resources are provided in support of risk analysis, with both qualitative and quantitative tools and techniques available.					
		D+R: Risk analysis outputs are compiled and disseminated in a format that supports decision making.					
	LEVEL 2–Repeatable The organization can repeat previous actions but cannot necessarily define those actions.	**Scope:** A basic risk analysis process is in place, following basic regulatory requirements and addressing "mission critical" risks, but with no internal requirement and no appetite for going beyond what is required by regulation.					
		Integration: Risk analysis is reactive, often initiated in response to an event or situation, or to justify decisions.					
		FB/OL: Risk analysis is largely rule based, regardless of context; or mirrors past projects, with limited organizational ability to plan ahead or learn from experience.					
		Resources: There is a limited range of tools and techniques available for and applied to risk analysis.					
	LEVEL 1–Initial	There is no formal process in place for risk analysis. An ad hoc approach exists. There is limited awareness of regulatory requirements for risk analysis.					

Process (Description)	Maturity level	Five maturity levels from learner to best practice*	Fully agree	Generally agree	Partially agree	Disagree	Evidence (Please use separate sheet if necessary)
Risk-based decision making and review The process by which the organization selects and develops its responses to risks. The output from this process is the input to risk response. Key tasks include 1. Selecting risk-response strategies based on output from risk analysis (e.g., avoid, retain, reduce, transfer, exploit). 2. Developing selected response strategies according to context and situation (e.g., a selected strategy of reducing currency risk may be developed into a complex hedging strategy). 3. Establishing indicators to track the progress and effectiveness of risk-response strategies. 4. Reviewing risk-response strategies with regard to risk-monitoring output.	LEVEL 5–Optimized The organization is "best practice," capable of learning and adapting. It not only uses experience to correct problems but also changes the nature of the way it operates.	**Scope:** Risk-based decision making proactively addresses all types of risk throughout all areas of the organization It extends to include the risk-based consideration of business opportunities and is often informed by formal cost–benefit analysis (e.g., for critical decisions).					
		Scope: The organization understands how decisions about risk will influence organizational performance (e.g., linking the organizational risk profile to corporate performance) and develops risk-response strategies with reference to broader organizational objectives.					
		Integration: Risk-based decision making is a continual, explicit component of all business processes, forming part of the organizational culture.					
		FB/OL: Feedback (from the risk-monitoring process) on the evolution of risks is actively used to revise and adapt decision making, and to explicitly question the principles and values that underlay "problem" decisions.					
		FB/OL: Risk-response strategies/risk-based decisions are proactively modified to adapt to internal and external changes and influences on the business.					
		Competence: The organization is able to go beyond the use of risk analysis outputs for answers (e.g., it can intuitively develop risk-response strategies where data is deficient), and considers the adaptation of organizational structures and practices to address the root causes of risk (i.e., changing the nature of its operations as part of its response to risk).					
	LEVEL 4–Managed The organization can control what it does in the way of processes. It determines requirements and ensures that these are met through feedback.	**Scope:** Risk-based decision making is actively practiced in accordance with risk policies and guidelines at all levels of the company and across all functional boundaries of the business.					
		Scope: Risk-based decisions are made with reference to the business unit and organizational risk profile, but without subverting functional priorities.					
		Integration: Risk-based decision making is automatically initiated as part of core business processes.					
		V+V: Risk-response strategies have measurable performance indicators (e.g., triggers and indicators established to highlight the need for revision), with systems in place to track progress and validate their expediency (these mechanisms are informed by the risk-monitoring process).					
		FB/OL: Feedback (from the risk-monitoring process) on the evolution of risks is actively used to revise and adapt decision making.					
		SE: Risk-based decision-making responsibilities generally reside within affected disciplines, and cross-functional stakeholders work together to define and implement an integrated approach to risk-based decision making, with input from key external stakeholders (e.g., regulatory bodies), where appropriate.					
		Competence: Risk analysis outputs are integral to risk-based decision making, and the organization can develop an extended range of effective risk-response strategies with reference to the life-cycle implications of risks.					
	LEVEL 3–Defined The organization can evaluate what it does and how it goes about it.	**Scope:** Risk-based decision making is defined, documented, and practiced with reference to risk policies and guidelines, addressing a broad range of risks throughout core business areas.					
		Scope: Risk-based decisions are made with reference to traditional functional priorities (e.g., in engineering, finance).					
		Integration: Procedures are in place to initiate risk-based decision making.					
		V+V: Basic mechanisms are in place to track whether risk-response strategies are meeting their objectives, largely reliant on lagging indicators (informed by the risk-monitoring process). The expertise for validation is generally lacking.					
		FB/OL: Risk-based decisions/risk-response strategies are reviewed and modified on an event-driven basis.					
		SE: Risk-based decision-making responsibilities generally reside within one group (e.g., risk management committee), with limited cross-functional or external consultation.					
		Competence: The organization understands how to use risk analysis outputs to inform risk-based decision making and selects from a broad range of risk-response strategies.					
	LEVEL 2–Repeatable The organization can repeat previous actions but cannot necessarily define those actions.	**Scope:** A basic process is in place for risk-based decision making, focused on repeating previously successful activities in order to manage critical risks, rather than on the explicit consideration of risk analysis outputs.					
		Scope: Risk-based decisions are largely made with reference to regulatory requirements, often isolated from business considerations and context.					
		Integration: Risk-based decision making is reactive, often initiated in response to an event or situation.					
		FB/OL: Risk-based decision making is largely rule based, regardless of context, or mirrors past projects, with limited organizational ability to plan ahead or learn from experience.					
	LEVEL 1–Initial	There is no structured approach established for risk-based decision making, with decisions made on an ad hoc basis. There is limited cognizance of risk analysis outputs.					

Process (Description)	*Maturity level*	*Five maturity levels from learner to best practice**	Fully agree	Generally agree	Partially agree	Disagree	Evidence *(Please use separate sheet if necessary)*
Risk response Risk response is the process of implementing risk-response strategies, such as the transference, reduction, avoidance, exploitation, and retention of risk. Key tasks include 1. Defining roles, responsibilities, and timescales for implementing risk-response strategies. 2. Prioritizing and allocate resources for the implementation of risk-response strategies. 3. Implementing risk-response strategies.	LEVEL 5–Optimized The organization is "best practice," capable of learning and adapting. It not only uses experience to correct problems but also changes the nature of the way it operates.	**Scope:** The organization is regarded as best practice in risk response at both the national and international level, with comprehensive and proactive response to all types of risk throughout all areas of the organization.					
		Integration: Risk response is a continual, explicit component of all business processes, forming part of the organizational culture.					
		V+V: Management continually establishes measurable targets for improving risk responsiveness, with systems in place to verify their achievement and validate the means through which they are pursued.					
		FB/OL: The organization uses feedback to link the efficacy of its risk-response efforts to business unit/organizational performance (e.g., by linking measures of residual risk to performance), acting as a spur for continual improvements in risk response.					
		FB/OL: Feedback is actively reviewed to improve the implementation of risk-response strategies, which further enables the adaptation of organizational structures and practices to optimize its ability to respond to risk.					
		FB/OL: The methods and techniques used in implementing risk-response strategies are proactively modified to adapt to internal and external changes and influences about the business.					
		Competence: A high level of competence exists among all staff, including management, involved in implementing risk-response strategies.					
		Resources: Resources for implementing risk-response strategies are sufficient and are optimally and flexibly allocated in response to need.					
	LEVEL 4–Managed The organization can control what it does in the way of processes. It determines requirements and ensures that these are met through feedback.	**Scope:** A controlled process is in place containing detailed criteria, methods, and guidelines to manage the implementation of risk-response strategies at all levels of the company and across all functional boundaries of the business.					
		Integration: Risk-response processes are automatically initiated as part of the core business processes.					
		V+V: Systems are in place to verify that risk-response strategies are efficiently implemented and to validate the expediency of implementation techniques.					
		FB/OL: Feedback is actively applied to improve the implementation of risk-response strategies (e.g., gaps are identified and risk-response techniques are improved in response to feedback).					
		SE: Risk-response processes generally reside within affected disciplines, and internal and external stakeholders work together to define and implement an integrated approach to risk response, capitalizing on synergies and collective knowledge (e.g., common-cause risks are addressed in a coordinated fashion, local industry and agriculture play their part in risk response, etc.).					
		Competence: Most involved staff exhibit a good level of competence in the implementation of risk-response strategies and have access to support from internal or external expert risk practitioners.					
		Resources: Sufficient resources are provided in support of risk response.					
	LEVEL 3–Defined The organization can evaluate what it does and how it goes about it.	**Scope:** A defined, documented process is in place containing criteria, methods, and guidelines for implementing risk-response strategies, addressing a broad range of risks throughout core business areas.					
		Integration: Procedures are in place to initiate risk-response processes.					
		V+V: Basic mechanisms are in place to verify that risk-response strategies are implemented as required, largely reliant on lagging indicators. The expertise for validation is generally lacking.					
		FB/OL: The implementation of risk-response strategies are reviewed and modified on an event-driven basis (e.g., in response to failures).					
		SE: Risk-response processes generally reside within the responsible unit, with limited cross-functional or external consultation.					
		Competence: Competency in the implementation of risk-response strategies resides only within the responsible unit.					
		Resources: Adequate resources are provided for the implementation of risk-response strategies.					
	LEVEL 2–Repeatable The organization can repeat previous actions but cannot necessarily define those actions.	**Scope:** A basic process is in place for the implementation of risk-response strategies, following basic regulatory requirements and addressing "mission critical" risks, but with no internal requirement and no appetite to go beyond what is required by regulation.					
		Integration: Risk response is reactive, often initiated in response to an event or situation.					
		FB/OL: Risk response is largely rule based, regardless of context, or mirrors past projects, with limited organizational ability to plan ahead or learn from experience.					
	LEVEL 1–Initial	There is no formal process in place for implementing risk-response strategies. An ad-hoc approach exists. There is a dependence on individual heroics.					

Process (Description)	Maturity level	Five maturity levels from learner to best practice*	Fully agree	Generally agree	Partially agree	Disagree	Evidence (Please use separate sheet if necessary)
Risk monitoring and feedback In this process, risk data are acquired, compiled, and reported by those responsible. Monitoring activities are largely based on the indicators established to track the progress and effectiveness of risk-response strategies. Monitoring data is transmitted in support of (particularly) risk-based decision making. Key tasks include 1. Defining criteria for risk-monitoring activities (i.e., detailing when, what, and how to monitor). 2. Collecting and updating data relating to the evolution of watched and mitigated risks. 3. Compiling and analyzing data for watched and mitigated risks to enable evaluation of the progress and effectiveness of risk-response strategies. NOTES: Watched risks are those that are accepted but deemed capable of becoming critical. Mitigated risks are those being actively managed.	LEVEL 5–Optimized The organization is "best practice," capable of learning and adapting. It not only uses experience to correct problems but also changes the nature of the way it operates.	**Scope:** The organization is regarded as best practice at both the national and international level in risk monitoring, with comprehensive and proactive monitoring of all types of watched and mitigated risks throughout all areas of the organization, using a "fit-for-purpose" range of tools and techniques.					
		Scope: The process extends to include the risk-based monitoring of business opportunities (e.g., tracking changes in the net present value of investment opportunities).					
		Integration: Risk monitoring is a continual, explicit component of all business processes, forming part of the organizational culture.					
		V+V: Management continually establishes measurable targets for improving the risk-monitoring process, with systems in place to verify their achievement and validate the means through which they are pursued.					
		FB/OL: The organization uses feedback to link trends in the evolution of the (aggregate) organizational risk profile to both organizational behavior (i.e., the ability to source organizational practices that are influencing the risk profile) and organizational performance (i.e., the ability to show how variations in the aggregate risk profile influence performance).					
		FB/OL: Feedback is actively used to improve both the philosophy and execution of risk monitoring, and further enables the adaptation of organizational structures and practices to optimize its ability to perform risk monitoring.					
		Competence: A high level of competence exists among all staff, including management, involved in risk monitoring.					
		Resources: Resources for risk monitoring are sufficient and are optimally and flexibly allocated in response to need.					
		Resources: The organization is active in researching and developing improved tools and techniques to enhance risk monitoring, with appropriate resources available.					
	LEVEL 4–Managed The organization can control what it does in the way of processes. It determines requirements and ensures that these are met through feedback.	**Scope:** A controlled process is in place containing detailed criteria, methods, and guidelines to manage the monitoring of watched and mitigated risks at all levels of the company and across all functional boundaries of the business.					
		Scope: The organization monitors changes in the (aggregate) risk profile at the organizational and business unit levels.					
		Integration: Risk-monitoring processes are automatically initiated as part of core business processes.					
		V+V: Systems are in place to verify the efficiency of risk-monitoring activities (i.e., the organization tracks monitoring activities to ensure they are performed to the required standard) and validate their expediency.					
		FB/OL: Feedback is actively used to improve the execution of risk-monitoring processes (e.g., gaps are identified and risk-monitoring techniques are improved in response to feedback).					
		SE: Risk-monitoring processes generally reside within affected disciplines, and stakeholders work together to define and implement an integrated approach to risk monitoring, capitalizing on synergies and collective knowledge (e.g., related risks are addressed in a coordinated fashion, water quality regulators provide input, etc.).					
		Competence: Most involved staff exhibit a good level of competence in applying risk-monitoring techniques and have access to support from internal or external expert risk practitioners.					
		Resources: Sufficient resources are provided in support of risk monitoring, a portion of which is made available for risk-monitoring R&D.					
		D+R: Risk status updates (e.g., "red," "green," and "amber") are compiled and disseminated in a clear, concise, and actionable format that supports real-time decision making, and their reporting is co-ordinated with other risk reporting mechanisms.					
	LEVEL 3–Defined The organization can evaluate what it does and how it goes about it.	**Scope:** A defined, documented process is in place containing criteria, methods and guidelines for the monitoring of a broad range of watched and mitigated risks throughout core business areas. The organization is conversant with and goes beyond regulatory requirements for risk monitoring.					
		Integration: Procedures are in place to initiate risk-monitoring processes.					
		V+V: Basic mechanisms are in place to verify that risk-monitoring activities are performed as required, largely reliant on lagging indicators. The expertise for validation is generally lacking.					
		FB/OL: Risk-monitoring activities are reviewed and modified on an event-driven basis.					
		SE: Risk-monitoring processes generally reside within the responsible unit, with limited cross-functional or external consultation.					
		Competence: Detailed knowledge of risk-monitoring techniques resides only within the responsible unit.					
		Resources: Adequate resources are provided in support of risk monitoring.					
		D+R: Risk status updates are compiled and disseminated in support of risk-based decision making.					
	LEVEL 2–Repeatable The organization can repeat previous actions but cannot necessarily define those actions.	**Scope:** Basic processes are in place to monitor watched and mitigated risks, following basic regulatory requirements and addressing "mission critical" risks, but with no internal requirement and no appetite to go beyond what is required by regulation.					
		Integration: Risk monitoring is reactive, often initiated in response to an event or situation.					
		FB/OL: Risk monitoring is largely rule based, regardless of context, or mirrors past projects, with limited organizational ability to plan ahead or learn from experience.					
		Resources: A limited range of tools and techniques exists that are available and applied to risk monitoring.					
	LEVEL 1–Initial	There are no formal processes in place for risk monitoring. An ad hoc approach exists. There is limited awareness of regulatory requirements for risk monitoring.					

Process (Description)	Maturity level	Five maturity levels from learner to best practice*	Fully agree	Generally agree	Partially agree	Disagree	Evidence (Please use separate sheet if necessary)
Integrating risk management This process involves the integration of risk management process interfaces, the cross-functional integration of risk management throughout the organization, and the integration of risk management with broader business operations. Key tasks include 1. Defining and integrating the risk management process interfaces (e.g., ensure that the outputs of the risk-monitoring process are sufficient to inform a review of risk-response strategies). Techniques for this purpose include technical exchange, peer reviews, etc. 2. Defining and implementing the desired level of cross-functional integration for risk management (i.e., the extent to which risk should be managed with respect to functional vs. organizational considerations). Integration is achieved through the establishment and utilization of methods for cross-functional coordination, solution building, conflict resolution, etc. 3. Integrating the process of risk management within broader business operations (e.g., through establishing initiation procedures, reward and accountability mechanisms, *etc.*).	LEVEL 5–Optimized The organization is "best practice," capable of learning and adapting. It not only uses experience to correct problems but also changes the nature of the way it operates.	**Scope of functional integration:** Management has defined and implemented a strategy for integrating risk management enterprise-wide, focused on integrating the management of risk toward broader corporate objectives (without subverting functional priorities). This is enabled by flexible management and organizational structures and an understanding of how the organizational risk profile influences corporate performance.					
		Integration within business: Risk management processes are continual, explicit components of all business processes, forming part of the organizational culture.					
		Integration within business: Company culture, policies, and reward and accountability systems promote the management of risk as "part of everyone's job."					
		V+V of interface integration: Verification and validation systems enable continuous improvement of risk management, such that changes to individual risk management processes have minimal impact on the risk management processes with which they interface.					
		FB/OL: Feedback is actively used to improve the integration process and further enables changes to organizational structures and practices to optimize the integration of risk management throughout the business.					
		Competence: All involved staff, including management, are fully cognizant of the methods and implications of integrating risk management.					
		Resources: Resources for integrating risk management are sufficient and are optimally and flexibly allocated in response to need.					
	LEVEL 4–Managed The organization can control what it does in the way of processes. It determines requirements and ensures that these are met through feedback.	**Scope of functional integration:** Management has defined and implemented a strategy for integrating risk management at all levels of the organization and across all functional boundaries of the business, focused on integrating the management of risk toward business unit and organizational risk profiles (without subverting functional priorities).					
		Integration within business: Risk management processes are automatically initiated as part of core business processes.					
		V+V of interface integration: Risk management process interfaces are understood and controlled, with procedures in place to coordinate them (e.g., technical exchange, reviewing inputs from each other, followup mechanisms for ensuring that appropriate responses are made in light of changes to monitored risks, etc.) and systems in place to verify and validate their coordination.					
		FB/OL: Senior management reviews the integration of risk management on a periodic and event-driven basis, modifying established strategy based on feedback from the risk management committee, corporate experiences, regulatory input, etc.					
		SE: Stakeholders work together to define and implement the integration of risk management throughout the organization (e.g., extensive cross-functional cooperation, engagement of regulatory bodies, etc.).					
		Competence: The most involved staff are aware of the methods and implications of integrating risk management.					
		Resources: Sufficient resources are provided in support of integrating risk management.					
	LEVEL 3–Defined The organization can evaluate what it does and how it goes about it.	**Scope of functional integration:** Management has defined and implemented a strategy for integrating risk management throughout core business areas, focused on integrating the management of risk toward functional priorities (e.g., engineering, finance).					
		Integration within business: Procedures are in place to initiate risk management processes within core business processes.					
		V+V of interface integration: Risk management process interfaces are identified and defined, with guidelines describing how they should interact (e.g., detailing how, when, and to whom risk analysis outputs should be disseminated), and basic mechanisms are in place to verify their interactions. The expertise for validation is generally lacking.					
		FB/OL: The integration of risk management is reviewed and modified on an event-driven basis.					
		SE: Responsibility for integrating risk management generally resides with one group (e.g., risk management committee).					
		Competence: Detailed knowledge of the methods and implications of integrating risk management resides only within the responsible group.					
		Resources: Adequate resources are provided in support of integrating risk management.					
	LEVEL 2–Repeatable The organization can repeat previous actions but cannot necessarily define those actions.	**Scope of functional integration:** Risks are largely managed with reference to basic regulatory requirements, often in isolation of functional considerations and context.					
		Integration within business: Risk management processes are reactive, often initiated in response to events or situations.					
		V+V of interface integration: Attempts to coordinate risk management process interfaces are initiated only when dependency failures are encountered (e.g., when risk-monitoring data is found insufficient to inform the review of risk-response strategies).					
	LEVEL 1–Initial	There are no formal efforts are made to integrate risk management. An ad hoc approach exists. There is limited cognizance of integrated risk management.					

Process (Description)	Maturity level	Five maturity levels from learner to best practice*	Fully agree	Generally agree	Partially agree	Disagree	Evidence (Please use separate sheet if necessary)
Supply chain risk management This is a process in which both product and service supply risks are addressed. Product supply risk management is the process that involves improving the way utilities obtain the raw components required to develop a product. Service supply risk management is the process that involves managing services provided by other organizations throughout the supply chain (e.g., outsourcing agreements). Managing the risks associated with supplier organizations is critical to maintaining an efficient and reliable service to customers. This is important because of the inherently extended nature of water utility supply chains and the increasing utilization of the outsourcing option. Key tasks include 1. Identifying and defining the risk interfaces between the organization and suppliers of services and products. 2. Establishing (risk-based) prequalification, selection and retention criteria for work performed by contractors, suppliers, and others. 3. Establishing (risk-based) deliverables and performance standards for suppliers of products and services (e.g., requirements for reliability, safety and technical competence). 4. Communicating the benefits of active risk management processes to supplier organizations.	LEVEL 5–Optimized The organization is "best practice," capable of learning and adapting. It not only uses experience to correct problems but also changes the nature of the way it operates.	**Scope:** Supply chain risk management encompasses the full risk implications of all supplier–organization interfaces, addressing product/service risk from concept to delivery.					
		Integration: Risk management is seamlessly integrated throughout the supply chain (i.e., the organization and its suppliers manage risk as a team, using feedback to improve their risk management processes).					
		V+V: Management continually establishes measurable targets for improving supply chain risk management, with systems in place to verify their achievement and validate the means through which they are pursued.					
		FB/OL: Feedback is actively used to revise the nature of supplier–organization relationships and, if necessary, to adapt broader organizational structures and practices (e.g., bringing retailing back "in-house") to optimize the management of risk throughout the supply chain.					
		SE: Suppliers are actively supported and coached in their risk management efforts.					
		Competence: The organization is conversant with international practice in supply chain risk management and is active in improving this, with all involved staff, including management, fully cognizant of the methods and implications of managing supply chain risks.					
		Resources: The organization understands the extent to which different supplier–organization interfaces influence its overall business performance, enabling the optimal and flexible allocation of sufficient resources in support of supply chain risk management activities.					
		Resources: Risk-based performance targets for suppliers are incentivized, acting as a spur for continual improvements.					
	LEVEL 4–Managed The organization can control what it does in the way of processes. It determines requirements and ensures that these are met through feedback.	**Scope:** A controlled process is in place to identify, analyze, and manage the extended risk implications of all supplier–organization interfaces, addressing technical, financial, delivery, and reputational risks.					
		Integration: Supply chain risk management is automatically initiated as part of core business processes.					
		V+V: Prequalification, selection, and retention criteria incorporate measurable risk-based performance standards for all suppliers (e.g., requirements for the reliability, safety, and technical competence of supplied products), and systems are in place to verify their adherence and validate their expediency.					
		FB/OL: Feedback on suppliers' risk-based performance is evaluated in real time with respect to agreed standards (though continuous improvement is not required) and is actively used to improve supply chain risk management (e.g., failures are responded to by increasing technical exchange with problem suppliers or altering selection criteria, etc.).					
		SE: Cooperative relationships are established throughout the supply chain, encouraging suppliers to work *with* the utility in managing risk for mutual benefit. All suppliers are encouraged to use an approved, tailored version of the organization's defined risk management process.					
		Competence: The organization is conversant with sectoral practice in supply chain risk management and makes efforts to exceed this, with most involved staff aware of the methods and implications of managing supply chain risks.					
		Resources: Sufficient resources are provided in support of supply chain risk management.					
	LEVEL 3–Defined The organization can evaluate what it does and how it goes about it.	**Scope:** A defined, documented process is in place to identify and evaluate the broad risk implications of core supplier–organization interfaces, addressing quality, costing, and timeliness risks.					
		Integration: Procedures are in place to initiate supply chain risk management activities.					
		V+V: Prequalification, selection, and retention criteria addressing risk are established for work performed by core suppliers, and basic mechanisms are in place to verify their adherence. The expertise for validation is generally lacking.					
		FB/OL: Supply chain risk management activities are reviewed and modified on an event-driven basis.					
		SE: Core suppliers are provided basic support in their risk management efforts (e.g., technical exchange and assistance) and are encouraged to implement a formal risk management process.					
		Competence: Detailed knowledge of the methods and implications of supply chain risk management resides only within the responsible unit.					
		Resources: Adequate resources are provided in support of supply chain risk management.					
	LEVEL 2–Repeatable The organization can repeat previous actions but cannot necessarily define those actions.	**Scope:** Basic procedures are in place for supply chain risk management, following basic regulatory requirements and addressing "mission critical" risks, but with no internal requirement and no appetite to go beyond what is required by regulation.					
		Integration: Supply chain risk management activities are reactive, often initiated in response to key failures throughout the supply chain.					
		FB/OL: Supply chain risk management activities are based on standards and previous experience, regardless of applicability or currency, with limited organizational ability to plan ahead or learn from experience (i.e., the organization manages risk by working with accredited suppliers and those who have performed well in the past).					
	LEVEL 1–Initial	There is no formal process in place to manage supply chain risks. An ad hoc approach exists. Reliance is entirely upon suppliers to manage risk.					

Process (Description)	Maturity level	Five maturity levels from learner to best practice*	Fully agree	Generally agree	Partially agree	Disagree	Evidence (Please use separate sheet if necessary)
Change management This involves managing the risk that is inherent to change. The purpose of change management is to identify and manage the risk implications of business (e.g., business process re-engineering) and technical (e.g., changes in design or technology) change. A dedicated change management process is particularly perceptive in the current operating climate, given the wide range of external influences (e.g., globalization, market restructuring) serving to fundamentally alter the context in which utilities conduct their business. Key tasks include 1. Defining and implementing an organizational strategy for the management of technical and business change. 2. Establishing (risk-based) criteria for acting upon change opportunities. 3. Identifying and defining the risk implications of changes in the operating, regulatory, and market environment (e.g., novel treatment technologies, alterations in compliance levels, evolving market structures, emerging competition, etc.). 4. Developing and executing plans to manage the risk implications of technical and business change.	LEVEL 5–Optimized The organization is "best practice," capable of learning and adapting. It not only uses experience to correct problems but also changes the nature of the way it operates.	***Integration:*** *Change management is a continual, explicit component of all business processes, forming part of the organizational culture.*					
		V+V: Management continually establishes measurable targets for improving the change management process, with systems in place to verify their achievement and validate the means through which they are pursued.					
		FB/OL: The organization proactively responds to significant external changes (e.g., market restructuring, changes in the regulatory regime, etc.), often adapting organizational structures and practices to pre-empt their risk implications.					
		FB/OL: Feedback from change management programs and pilot tests is actively used to identify and disseminate best practices throughout the organization, adapting organizational structures and practices where necessary.					
		SE: Staff at all levels of the company are engaged in change management activities, creating an organizational culture that confronts and adapts to change.					
		Competence: The organization is conversant with international practice in change management and is active in improving this, with all involved staff, including management, fully cognizant of the methods and implications of change management.					
		Resources: The organization understands the extent to which various change management programs influence its overall business performance, enabling the optimal and flexible allocation of sufficient resources in support of change management.					
		Resources: Pilot tests are performed before core internal changes to determine feasibility and improvement opportunities, with appropriate resources available.					
	LEVEL 4–Managed The organization can control what it does in the way of processes. It determines requirements and ensures that these are met through feedback.	**Scope:** A controlled process is in place to identify, evaluate (with respect to risk), and manage all changes, however sourced, that could significantly alter the organizational risk profile.					
		Integration: Change management is automatically initiated as part of core business processes.					
		V+V: The objectives of change management programs are understood in measurable terms, and systems are in place to verify their achievement (e.g., tracking to ensure that the effect of altering treatment design on process reliability is as desired) and validate their expediency.					
		FB/OL: Feedback on the progress of change management programs is reviewed in real time with respect to set objectives, and the execution of the process evolves based on this information.					
		SE: Change management processes generally reside within the disciplines causing change (e.g., planning and scheduling, engineering) and are informed by cross-functional and external consultation.					
		SE: Key external stakeholders (i.e., those responsible for, or similarly affected by, change) are mutual partners in change management (e.g., the organization works with the water quality regulator to ensure that the outcome of water quality standard reviews is mutually acceptable).					
		Competence: The organization is conversant with sectoral practice in change management and makes efforts to exceed this, with most involved staff aware of the methods and implications of change management.					
		Resources: Sufficient resources are provided in support of change management activities.					
	LEVEL 3–Defined The organization can evaluate what it does and how it goes about it.	**Scope:** A defined, documented process is in place to identify and plan for the risk implications of organizational (e.g., re-engineering efforts), technical (e.g., alterations in design or manufacture), and regulatory (e.g., pricing reviews) change.					
		Integration: Procedures are in place to initiate change management activities.					
		V+V: Basic mechanisms are in place to verify that change management efforts are meeting their objectives, largely reliant on lagging indicators. The expertise for validation is generally lacking.					
		FB/OL: Change management programs are reviewed and modified on an event-driven basis.					
		SE: Processes to manage change generally reside within one unit (e.g., planning and scheduling), with limited cross-functional or external consultation.					
		Competence: Detailed knowledge of the methods and implications of change management resides only within the responsible unit.					
		Resources: Adequate resources are provided in support of change management activities.					
	LEVEL 2–Repeatable The organization can repeat previous actions but cannot necessarily define those actions.	**Scope:** Basic processes are in place to manage critical changes, although the organization cannot define the risk implications of change.					
		Integration: Change management processes are reactive, often initiated in response to an event or situation (i.e., the organization may be blind to change, responding only after the event).					
		FB/OL: Change management activities are retrospective and historical, regardless of applicability or currency, with limited organizational ability to plan ahead or learn from experience.					
	LEVEL 1–Initial	There are no formal processes in place to manage change. An ad hoc approach exists.					

Process (Description)	Maturity level	Five maturity levels from learner to best practice*	Fully agree	Generally agree	Partially agree	Disagree	Evidence (Please use separate sheet if necessary)
Education and training in risk management Education and training involves the development, maintenance, and enhancement of the skills and knowledge that enable staff to efficiently perform their risk management roles. Key tasks include 1. Defining education and training requirements for effective risk management (i.e., competency requirements). 2. Designing education programs and training vehicles to impart the required knowledge. 3. Tracking the progress and effectiveness of education and training in risk management. 4. Defining and implementing development opportunities for staff (i.e., education and training opportunities that extend beyond role requirements) to optimize risk management throughout the organization.	LEVEL 5–Optimized The organization is "best practice," capable of learning and adapting. It not only uses experience to correct problems but also changes the nature of the way it operates.	**Scope:** Education and training in risk management is continuous and extends beyond operational aspects (i.e., ensuring staff can perform current roles) to develop broader knowledge structures to optimize risk management throughout the organization (e.g., embedding a knowledge of how human factors and organizational structures and practices influence risk).					
		Integration: Education and training in risk management is a continual, explicit component of all business processes, forming part of the organizational culture.					
		V+V: Management continually establishes measurable targets for improving risk management education and training, with systems in place to verify their achievement and validate the means through which they are pursued.					
		FB/OL: Feedback is actively used to improve both the philosophy and execution of risk management education and training, and further enables the adaptation of organizational structures and practices to optimize its ability to educate and train staff about risk management.					
		FB/OL: Risk management education and training programs are proactively modified in response to internal and external changes and influences on the business.					
		Competence: The organization is conversant with international practice in risk management education and training and is active in improving this, with all involved staff (i.e., trainers and educators) exhibiting a high level of competence.					
		Resources: Resources for education and training in risk management are sufficient and are allocated optimally and flexibly in response to need.					
		Resources: Coaches are provided to support continuous development of individual or team competencies in risk management.					
		Resources: Risk management performance is incentivized to encourage continuous development of skills and knowledge.					
	LEVEL 4–Managed The organization can control what it does in the way of processes. It determines requirements and ensures that these are met through feedback.	**Scope:** A controlled process is in place to manage the education and training of all staff in risk management, addressing the full range of technical and managerial aspects and applied at all levels of the company and across all functional boundaries of the business.					
		Integration: Education and training in risk management is automatically initiated as part of core business processes.					
		V+V: The required skills and knowledge (competencies) for effective risk management are understood in measurable terms (e.g., translated into performance data), and systems are in place to verify their development/maintenance and validate the means through which they are developed/maintained.					
		FB/OL: Skills and knowledge developed at the individual, team, unit, and organizational level are evaluated against required competencies, and the execution of education and training programs is evolved based on this feedback.					
		SE: Stakeholders work together to define and implement an integrated approach to education and training throughout the organization.					
		Competence: The organization is conversant with sectoral practice in risk management education and training and makes efforts to exceed this, with most involved staff (i.e., trainers and educators) exhibiting a good level of competence.					
		Resources: Sufficient resources are provided in support of risk management education and training, with a wide range of formal and informal vehicles adopted to impart the required skills and knowledge. Staff have access to educational resources on risk management outside the formal program.					
	LEVEL 3–Defined The organization can evaluate what it does and how it goes about it.	**Scope:** A defined, documented process is in place to educate and train core staff in risk management, addressing a broad range of technical and managerial issues and applied throughout core business areas.					
		Integration: Procedures are in place to initiate education and training in risk management.					
		V+V: Basic mechanisms are in place to verify that the skills and knowledge (competencies) required for effective risk management are developed/maintained through education and training, largely reliant on lagging indicators. The expertise for validation is generally lacking.					
		FB/OL: Risk management education and training activities are reviewed and modified on an event-driven basis.					
		SE: Responsibility for educating and training staff in risk management resides within one group.					
		Competence: Detailed knowledge of techniques and methods for educating and training staff in risk management resides only with senior-level involved staff (i.e., experienced trainers and educators).					
		Resources: Adequate resources are provided in support of risk management education and training, with both formal and informal vehicles adopted.					
	LEVEL 2–Repeatable The organization can repeat previous actions but cannot necessarily define those actions.	**Scope:** Education and training in risk management exists but is restricted to ensuring that critical staff (e.g., finance, reliability engineering) have the required technical skills and knowledge to perform repeat functions in their current risk management roles.					
		Integration: Education and training in risk management is reactive, often initiated in response to major failures in risk management.					
		FB/OL: Education and training programs are retrospective and historical, regardless of applicability or currency, with limited organizational ability to plan ahead or learn from experience.					
		Resources: Risk management education and training requirements are largely undefined and are largely reliant on informal vehicles (e.g., on-the-job training).					
	LEVEL 1–Initial	There are no processes in place to develop or maintain risk management skills and knowledge. An ad hoc approach exists. There is limited cognizance of required competencies for effective risk management.					

Process (Description)	Maturity level	Five maturity levels from learner to best practice*	Fully agree	Generally agree	Partially agree	Disagree	Evidence (Please use separate sheet if necessary)
Risk knowledge management Risk knowledge management includes the gathering and organization of risk information, constant refinement through analysis and collaboration, and dissemination in such a manner as to ensure its effective and consistent use in support of utility decision making. Key tasks include 1. Defining data/information requirements for effective risk knowledge management. 2. Designing, developing, and implementing risk knowledge management systems and infrastructure to capture, compile, and analyze the required data/information.	LEVEL 5–Optimized The organization is "best practice," capable of learning and adapting. It not only uses experience to correct problems but also changes the nature of the way it operates.	**Scope:** Organization is regarded as best practice in risk knowledge management, possessing world-class systems and infrastructure for recognizing, creating, transforming, and distributing risk knowledge, with the intention of satisfying operational risk management requirements and developing organizational "risk intelligence."					
		Integration: Risk knowledge management is a continual, explicit component of all business processes, forming part of the organizational culture.					
		V+V: Management continually establishes measurable targets for improving risk knowledge management, with systems in place to verify their achievement and validate the means through which they are pursued.					
		FB/OL: Feedback is actively used to improve both the philosophy and execution of risk knowledge management processes, and further enables the adaptation of organizational structures and practices (e.g., changing incentive structures to encourage knowledge sharing and collaboration between traditionally "competing" departments) to optimize risk knowledge management throughout the business.					
		FB/OL: Risk information systems and supporting infrastructure are proactively modified to meet changing requirements and situations (e.g., as objectives and related risks change).					
		Competence: The organization is conversant with international practice in risk knowledge management and is active in improving this, with all involved staff, including management, possessing a high level of competence in the methods and techniques of risk knowledge management.					
		Resources: Resources for risk knowledge management are sufficient and are optimally and flexibly allocated in response to need.					
		Resources: The organization is active in researching and developing improved systems and infrastructure for risk knowledge management, with appropriate resources available.					
	LEVEL 4–Managed The organization can control what it does in the way of processes. It determines requirements and ensures that these are met through feedback.	**Scope:** The gathering, storage, analysis, and management of risk knowledge is automated and networked within all levels of the organization and across all functional boundaries of the business. The required granularity, volume, and accuracy of risk information are defined.					
		Integration: Risk knowledge management processes are automatically initiated as part of core business processes.					
		V+V: Systems are in place to verify the efficiency of risk knowledge management processes (e.g., the organization tracks the timeliness, relevancy, breadth, and depth of risk information collected and disseminated with reference to defined requirements) and validate their design (e.g., questioning whether the information infrastructure is best suited to meet process requirements).					
		FB/OL: Feedback is actively used to improve the execution of risk knowledge management processes (e.g., failures identified and information infrastructure improved in response).					
		SE: Internal and external stakeholders work together to define and implement an integrated approach to risk knowledge management (e.g., the organization mutually exchanges risk information with regulators).					
		Competence: The organization is conversant with sectoral practice in risk knowledge management and makes efforts to exceed this, with most involved staff possessing a high level of competence in the methods and techniques of risk knowledge management.					
		Resources: Sufficient resources are provided in support of risk knowledge management, a portion of which is made available for R&D in risk knowledge management.					
		D+R: Risk information (e.g., risk analysis outputs, risk status updates) is compiled and disseminated in a clear, concise, and actionable format that supports real-time decision making, and risk reporting mechanisms are effectively coordinated.					
	LEVEL 3–Defined The organization can evaluate what it does and how it goes about it.	**Scope:** A defined, documented, and largely automated process for risk knowledge management is in place, providing a framework for the gathering, storage, analysis, and distribution of risk information and is applied throughout core business areas, with the intention of satisfying operational risk management requirements.					
		Integration: Procedures are in place to initiate risk knowledge management processes.					
		V+V: Basic mechanisms are in place to verify that the organization is meeting its information requirements for risk management, largely reliant on lagging indicators. The expertise for validation is generally lacking.					
		FB/OL: Risk knowledge management processes are reviewed and modified on an event-driven basis.					
		SE: Risk information transfer is generally restricted within divisions or groups, with limited communication on risk across organizational boundaries.					
		Competence: Detailed knowledge of the methods and techniques of risk knowledge management resides only within the responsible unit.					
		Resources: Adequate resources are provided in support of risk knowledge management.					
		D+R: Risk information (e.g., risk analysis outputs, risk status updates) is compiled and disseminated in a format that supports decision making.					
	LEVEL 2–Repeatable The organization can repeat previous actions but cannot necessarily define those actions.	**Scope:** Basic processes have been established for the collection and storage of risk information, restricted to satisfying basic regulatory requirements and informing the management of "mission critical" risks.					
		Integration: Risk knowledge management processes are reactive, often initiated in response to an event or a situation.					
		FB/OL: Risk knowledge management systems and infrastructure are retrospective and historical, regardless of applicability or currency, with limited organizational ability to plan ahead or learn from experience.					
	LEVEL 1–Initial	There are no formal processes in place to manage risk data and information. An ad hoc approach exists. There is limited awareness of information requirements for risk management.					

*NOTES:
Scope: The extent or range of risk management throughout the organization.
Integration: The integration of risk management processes within broader business processes and operations.
V+V: Verification and validation—Verification is ensuring that the model, procedure, process, etc., is being correctly performed, and validation is ensuring that the correct model, procedure, process, etc., is being performed.
FB/OL: Feedback and organizational learning—The manner in which feedback, both internally and externally sourced, is collected and used to question and revise processes.
SE: Stakeholder engagement—The process of engaging stakeholders, both internal and external, with the purpose of optimizing risk management activities.
Competence: The organizational qualities and abilities, both implicit and explicit, that influence process performance.
Resources: People, costs, and tools.
D+R: Documentation and reporting—The process of documenting and reporting risk information.

REFERENCES

Aabo, T., J. Fraser, and B. Simkins. 2004. The Rise and Transformation of the Chief Risk Officer: A Success Story on Enterprise Risk Management, Social Science Research Network Electronic Library [Online]. New York: Social Science Research Network. Available: <http://ssrn.com/abstract=622744>. [cited November 14, 2006]

Abell, P., and M. Askey. 2001. Process Risk Assessment: Water Supply Systems. In *Risk Assessment for Environmental Professionals*. Edited by S.J. Pollard and J.A. Guy. Suffolk, England: Lavenham Press.

AIRMIC, ALARM, and IRM (Association of Insurance and Risk Managers, National Forum for Risk Management in the Public Sector, and Institute of Risk Management). 2002. *A Risk Management Standard* [Online]. London: AIRMIC. Available: <http://www.airmic.com>. [cited November 14, 2006]

Allison, I., and P. Chadwick. 2003. Balancing Risk in Asset Management Planning for Water Resources and Capital Maintenance. In *Proc. of the ICE/CIWEM Water Environment Conference: Risk and Reward in Asset Management Delivery—Who Is Best Prepared for the Challenges Ahead*. London: Institution of Civil Engineers/Chartered Institution of Water and Environmental Management.

Amatetti, E.J. 1994. Managing the Financial Condition of a Utility. *Journal AWWA*, 86(4):176–187.

Argyris, C., and D. Schön. 1978. *Organizational Learning: A Theory of Action Perspective*, Reading, Mass.: Addison Wesley.

ASCE (American Society of Civil Engineers). 2001. Report Card for America's Infrastructure [Online]. Washington, D.C.: ASCE. Available: <http://www.asce.org>. [cited November 14, 2006]

AS/NZS (Australian/New Zealand Standard). 1995. *Risk Management. AS/NZS 4360:1995*. Strathfield, New South Wales: Standards Association of Australia.

AS/NZS (Australian/New Zealand Standard). 1999. *Risk Management. AS/NZS 4360:1999*. Strathfield, New South Wales: Standards Association of Australia.

AS/NZS (Australian/New Zealand Standard). 2004. *Risk Management. AS/NZS 4360:2004*. Strathfield, New South Wales: Standards Association of Australia.

AWWA (American Water Works Association). 2003. Blackout Highlights Water Security Issues [Online]. *E-Mainstream*, 47(3). Available: <http://www.awwa.org>. [cited November 14, 2006]

AwwaRF (Awwa Research Foundation). 1997. *The Utility Business Architecture: Designing for Change*. Denver, Colo.: AwwaRF and AWWA.

AwwaRF (Awwa Research Foundation). 2004. *Risk Communication for Emerging Contaminants*. Denver, Colo.: AwwaRF.

AwwaRF (Awwa Research Foundation). 2005. *The Value of Water: Concepts, Estimates and Applications for Water Managers*. Denver, Colo.: AwwaRF.

AwwaRF (Awwa Research Foundation). 2006a. *Application of HACCP for Distribution System Protection*. Denver, Colo.: AwwaRF.

AwwaRF (Awwa Research Foundation). 2006b. *Asset Management Planning and Reporting Options for Water Utilities*. Denver, Colo.: AwwaRF and AWWA.

AwwaRF (Awwa Research Foundation). 2006c. *Emergency Response and Recovery Planning for Drinking Water Systems: A Kit of Tools*. Denver, Colo.: AwwaRF and AWWA.

AwwaRF (Awwa Research Foundation). 2006d. *Estimating Health Risks From Infrastructure Failures*. Denver, Colo.: AwwaRF, AWWA, and IWA Publishing.

AwwaRF (Awwa Research Foundation). 2006e. *A Knowledge Management Approach to Drinking Water Utility Business*. Denver, Colo.: AwwaRF.

Badenoch, J. 1990. Cryptosporidium *in Water Supplies—Report of the Group of Experts*. London: Her Majesty's Stationery Office.

Booth, R., and J. Rogers. 2001. Using GIS Technology to Manage Infrastructure Capital Assets. *Journal AWWA*, 93(11):62.

Buckley, C.B. 2004. Communications. In *Water Contamination Emergencies: Can We Cope?* Edited by K.C. Thompson and J. Gray. Cambridge: Royal Society of Chemistry.

Cascos, G., J. Ciaccia, J. Tripp, J. Bella, S. Qussar, and M. Lacey. 2004. Emergency Backup Power Supply (Chat Room). *Journal AWWA*, 96(1):41–43.

Committee on Corporate Governance. 1998. *The Combined Code*. London: Gee Publishing.

COSO (Committee of Sponsoring Organizations of the Treadway Commission). 1992. *Internal Control—Integrated Framework* (Executive Summary and Evaluation Tools). 2 volumes. New York: American Institute of Certified Public Accountants.

COSO (Committee of Sponsoring Organizations of the Treadway Commission). 2004. *Enterprise Risk Management—Integrated Framework* (Executive Summary) [Online]. Available: <http://www.otpp.com>. [cited November 14, 2006]

Crosby, P.B. 1979. *Quality Is Free*. New York: McGraw-Hill.

Crosby, P.B. 1996. *Quality Is Still Free*. New York: McGraw-Hill.

CSA (Canadian Standards Association). 1997. *Risk Management: Guidelines for Decision Makers*. CAN/CSA-Q850-97. Mississauga, Ont.: CSA.

Cunliffe, D. 2003. Application of the Australian Risk Management Framework: A Public Health Authorities' Perspective. In *Proceedings of Water Safety—Risk Management Strategies for Drinking Water*. Berlin: *Umweltbundesamt* (Federal Environmental Agency).

Deere, D., and A. Davison. 1998. Safe Drinking Water—Are Food Guidelines the Answer? *Water*, 25:21–24.

Deere, D., M. Stevens, A. Davison, G. Helm, and A. Dufour. 2001. Management Strategies. In *Water Quality: Guidelines, Standards and Health*. Edited by L. Fewtrell and J. Bartram. London: IWA Publishing.

Deloach, J.W. 2000. Enterprise-Wide Risk Management—Strategies for Linking Risk and Opportunity. London: Pearson Education.

Doyle, G., and M. Grabinsky. 2003. Applying GIS to a Water Main Corrosion Study. *Journal AWWA*, 93(5):90.

enHealth Council 2002. *Environmental Health Risk Assessment: Guidelines for Assessing Human Health Risks From Environmental Hazards*. Perth, Australia: enHealth Council.

Environmental Data Services. 1999. Camelford: Questions Still Unanswered After Eleven Years [Online]. *The ENDS Report*, Issue 297. Available: <http://www.environmental-assistance.org>. [cited November 14, 2006]

Foster, J.A., and A.T. McDonald. 2000. Assessing Pollution Risks to Water Supply Intakes Using Geographic Information Systems (GIS). *Environmental Modelling and Software*, 15:225.

Fraser, J. 2005. Corporate Risk Management in the Utility Sector—Hydro One Inc. In *Proc. of the AwwaRF International Workshop: Risk Analysis Strategies for Better and More Credible Decision-Making*. Edited by S.J.T. Pollard, S.E. Hrudey, L. Reekie, and P.D. Hamilton. Denver, Colo.: AwwaRF and Cranfield University, U.K.

Fraser, P., J. Moultrie, and M. Gregory. 2002. The Use of Maturity Models/Grids as a Tool in Assessing Product Development Capability. Paper presented at IEEE International Engineering Management Conference: Managing Technology for the New Economy. St. John's College, Cambridge, August 18–20. New Jersey: Institute of Electrical and Electronics Engineers.

Fraser, R. 1998. *The Burncrooks Inquiry—Report on the Disruption to Public Water Supplies in the Area Served by Burncrooks Waterworks, December 1997* [Online]. The Scottish Office. Available: <http://www.scotland.gov.uk>. [cited November 14, 2006]

Grimsey, D., and M.K. Lewis. 2002. Evaluating the Risks of Public Private Partnerships for Infrastructure Projects. *International Journal of Project Management*, 20:107.

Haas, C.N., J.B. Rose, and C.P. Gerba. 1999. Risk Assessment Paradigms. In *Quantitative Microbial Risk Assessment*. New York: John Wiley and Sons.

Hrudey, S.E., and E.J. Hrudey. 2004. *Safe Drinking Water. Lessons From Recent Outbreaks in Affluent Nations*. London: IWA Publishing.

Hrudey, S.E., E.J. Hrudey, and S.J.T. Pollard. 2006. Risk Management for Assuring Safe Drinking Water. *Environment International*, 32:948–957.

Hrudey, S.E., P.M. Huck, P. Payment, R.W. Gillham, and E.J. Hrudey. 2002. Walkerton: Lessons Learned in Comparison With Waterborne Outbreaks in the Developed World. *Journal of Environmental Engineering and Science*, 1:397–407.

Huber, J.L. 2000. The Evolving Competitive Marketplace. *Journal AWWA*, 92(1):50–51.

Hurst, N.W. 1998. *Risk Assessment: The Human Dimension*. Cambridge: Royal Society of Chemistry.

IACCM (International Association for Contract and Commercial Management). 2003. *Organisational Maturity in Risk Management: The IACCM Business Risk Management Maturity Model (BRM3)* [Online]. Available: <http://www.risk-doctor.com/pdf-files/brm1202.pdf>. [cited November 14, 2006]

ICAEW (Institute of Chartered Accountants in England and Wales). 1999a. *Implementing Turnbull—A Boardroom Briefing* [Online]. London: ICAEW. Available: <http://www.icaew.co.uk>. [cited November 14, 2006]

ICAEW (Institute of Chartered Accountants in England and Wales). 1999b. *Internal Control: Guidance for Directors on the Combined Code* [Online]. London: ICAEW. Available: <http://www.icaew.co.uk>. [cited November 14, 2006]

ILSI (International Life Sciences Institute). 1996. A Conceptual Framework to Assess the Risks of Human Disease Following Exposure to Pathogens. *Risk Analysis*, 16(6):841–848.

ILSI (International Life Sciences Institute). 2000. *Revised Framework for Microbial Risk Assessment* [Online]. ILSI Risk Science Institute Workshop Report. Washington, D.C.: ILSI. Available: <http://www.ilsi.org>. [cited November 14, 2006]

ISO (International Organization for Standardization). 1996. *Environmental Management Systems—Specification With Guidance for Use*. ISO 14001:1996. Geneva: ISO.

ISO (International Organization for Standardization). 2000. *Quality Management Systems—Guidelines for Performance Improvement*. ISO 9004. London: British Standards Institution.

ISO/IEC (International Organization for Standardization/International Electrotechnical Commission). 2002. *Risk Management–Vocabulary–Guidelines for Use in Standards*. ISO/IEC Guide 73:2002. London: British Standards Institution.

IWA (International Water Association). 2004. *The Bonn Charter for Safe Drinking Water*. London: IWA Publishing.

Jansen, R.B. 1988. *Advanced Dam Engineering for Design, Construction and Rehabilitation.* New York: Van Nostrand Reinhold.

Jardine, C.G., S.E. Hrudey, J.H. Shortreed, D. Krewski, C. Frugal, L. Craig, and S. McColl. 2003. Review of Risk Management Frameworks for Environmental, Human Health and Occupational Health Risks. *Journal of Toxicology and Environment Health, Part B,* 6(6):569–718.

Kaplan, R.S., and D.P. Norton. 1996. *The Balanced Scorecard: Translating Strategy Into Action.* Boston: Harvard Business School Press.

Kaufman, M.M., and M. Wurtz. 1998. Small System Maintenance Management Using GIS. *Journal AWWA*, 90(8):70.

Kent, N.A., D. Treece, and M. Davies. 2003. Asset Management Through Risk Assessment for Welsh Water's Water Trunk Mains. In *Proc. of the ICE/CIWEM Conference: Risk and Reward in Asset Management Delivery—Who is Best Prepared for the Challenges Ahead?* London: Institution of Civil Engineers/Chartered Institution of Water and Environmental Management.

Kletz, T. 2001. *Learning From Accidents.* 3rd ed. Oxford: Elsevier.

Laing, R.D. 2002. *Report of the Commission of Inquiry Into Matters Relating to the Safety of the Public Drinking Water in the City of North Battleford, Saskatchewan.* Regina, Sask.: Queen's Printer.

Lam, J. 2003. *Enterprise Risk Management: From Incentives to Controls.* Hoboken, N.J.: John Wiley and Sons.

Lam, J. 2005. Best Practices in Developing Key Risk Indicators and ERM Reporting. An Executive White Paper. Wellesley, Mass.: James Lam and Associates.

Leiss, W. 2005. Risk Management: Coverup of Tool for Better Decisions. In *Proc. of the AwwaRF International Workshop: Risk Analysis Strategies for Better and More Credible Decision-Making.* Edited by S.J.T. Pollard, S.E. Hrudey, L. Reekie, and P.D. Hamilton. Denver, Colo.: AwwaRF and Cranfield University, U.K.

Leverett, S. 2003. Eating the Elephant. *Water*, 166:8–9.

Lifton, G., and P. Smeaton. 2003. Asset Risk Management in Scottish Water. In *Proc. of the ICE/CIWEM Water Environment Conference: Risk and Reward in Asset Management Delivery—Who Is Best Prepared for the Challenges Ahead?* London: Institution of Civil Engineers/Chartered Institution of Water and Environmental Management.

Lifton, G., and P. Smeaton. 2005. Risk and Asset Management—The Interface. In *Proc. of the AwwaRF International Workshop: Risk Analysis Strategies for Better and More Credible Decision-Making.* Denver, Colo.: AwwaRF and Cranfield University, U.K.

Lisle, J.T., and J.B. Rose. 1995. *Cryptosporidium* Contamination of Water in the USA and UK: A Mini-Review. *Journal of Water Supply: Research and Technology–Aqua*, 44(3):103–117.

Lloyd, R., and P. Abell. 2005. Risk Analysis Strategies for Better and More Credible Utility Decisions: Project Overview and Progress. In *Proc. of the AwwaRF International Workshop: Risk Analysis Strategies for Better and More Credible Decision-Making.* Edited by S.J.T. Pollard, S.E. Hrudey, L. Reekie, and P.D. Hamilton. Denver, Colo.: AwwaRF and Cranfield University, U.K.

Luehrman, T.A. 1997. What's It Worth? A General Manager's Guide to Valuation. *Harvard Business Review*, 75(3):132.

MacGillivray, B.H., P.D. Hamilton, J.E. Strutt, and S.J.T. Pollard. 2006a. Risk Analysis Strategies in the Water Utility Sector: An Inventory of Applications for Better and More Credible Decision Making. *Critical Reviews in Environment Science and Technology*, 36:85–139.

MacGillivray, B.H., J.V. Sharp, J.E. Strutt, P.D. Hamilton, and S.J.T. Pollard. 2006b. Benchmarking Risk Management Within the International Water Utility Sector. Part I: Design of a Capability Maturity Methodology. *Journal of Risk Research*. Forthcoming.

MacGillivray, B.H., J.V. Sharp, J.E. Strutt, P.D. Hamilton, and S.J.T. Pollard. 2006c. Benchmarking Risk Management Within the International Water Utility Sector. Part II: A Survey of Eight Water Utilities. *Journal of Risk Research*. Forthcoming.

Marshall, A. 2001. Corporate Governance and Risks Management: Obligations for Corporate Accountability. In *Proc. of the IBC Environmental Benchmarking and Reporting Conference*. London: IBC Conferences.

McClellan, P. 1998. *Sydney Water Inquiry*. New South Wales, Australia: Premier's Department.

McKay, D. 2005. A U.S. Utility Perspective: Louisville Water Co.—Somewhere on the Road to Enterprise Risk Management. In *Proc. of the AwwaRF International Workshop: Risk Analysis Strategies for Better and More Credible Decision-Making*. Denver, Colo.: AwwaRF and Cranfield University, U.K.

Means, E.G., L. Ospina, and R. Patrick. 2005. Ten Primary Trends and the Implications for Water Utilities. *Journal AWWA*, 97(7):64–77.

Miller, A.G. 2005. Risk Governance in a Water Utility. In *Proc. of the AwwaRF International Workshop: Risk Analysis Strategies for Better and More Credible Decision-Making*. Edited by S.J.T. Pollard, S.E. Hrudey, L. Reekie, and P.D. Hamilton. Denver, Colo.: AwwaRF and Cranfield University, U.K.

NAS (National Academy of Sciences). 1983. *Risk Assessment in the Federal Government: Managing the Process*. Washington, D.C.: National Academy Press.

NERAM (Network for Environmental Risk Assessment and Management). 2002. *Drinking Water Safety: A Total Quality Management Approach* [Online]. Waterloo, Ont.: NERAM. Available: <http://irr.uwaterloo.ca/pdf_files/waterFinal.pdf>. [cited November 14, 2006]

NERAM (Network for Environmental Risk Assessment and Management). 2003. *Basic Frameworks for Risk Management* [Online]. Waterloo, Ont.: NERAM. Available: <http://irr.uwaterloo.ca/pdf_files/basicFrameworkMar2003.pdf>. [cited November 14, 2006]

New Zealand Ministry of Health. 2001. *How to Prepare and Develop Public Health Risk Management Plans for Drinking-Water Supplies* [Online]. Wellington, New Zealand: Ministry of Health. Available: <http://www.moh.govt.nz>. [cited November 14, 2006]

NHMRC (National Health and Medical Research Council). 2001. *Framework for Management of Drinking Water Quality*. Canberra: NHMRC.

NHMRC (National Health and Medical Research Council). 2002. *Australian Drinking Water Guidelines—Draft*. Canberra: NHMRC.

NHMRC (National Health and Medical Research Council). 2004. *Australian Drinking Water Guidelines* [Online]. Canberra: NHMRC. Available: <http://www.gov.au/nhmrc/publications/synopses/eh19syn.htm>. [cited November 14, 2006]

O'Connor, D.R. 2002a. *The Events of May 2000 and Related Events*. Report of the Walkerton Inquiry (Part 1). Toronto, Ont.: Ontario Ministry of the Attorney General.

O'Connor, D.R. 2002b. *A Strategy for Safe Water*. Report of the Walkerton Inquiry (Part 2). Toronto, Ont.: Ontario Ministry of the Attorney General.

Parr, N.M., and N. Cullen. 1988. Risk Management and Reservoir Maintenance. *Journal of the Institution of Water and Environmental Management*, 2:587.

Patrick, R., and E.G. Means. 2005. Meeting Customer Expectations in a Fluid Environment. *Journal AWWA*, 97(9):56–61.

Paulk, M.C., M. Chrissis, and C.V. Weber. 1993. Capability Maturity Model for Software, Version 1.1. *IEEE Software,* 10(4):18–27.

Pollard, S., and G. Carroll. 2001. Recent Developments and New Directions. In *Risk Assessment for Environmental Professionals*. Edited by S. Pollard and J. Guy. Chartered Institution of Water and Environmental Management. Suffolk, England: Lavenham Press.

Pollard, S.J.T., S.E. Hrudey, L. Reekie, and P.D. Hamilton, eds. 2005. *Proc. of the AwwaRF International Workshop: Risk Analysis Strategies for Better and More Credible Decision-Making*. Denver, Colo.: AwwaRF and Cranfield University, U.K.

Pollard, S.J.T., B.H. MacGillivray, P.D. Hamilton, S.E. Hrudey, J.E. Strutt, J.V. Sharp, and L. Reekie. 2006a. Better Utility Decisions—Evaluating Organisational Maturity in Risk Management Within the Water Utility Sector. Paper presented at the *WEF/AWWA 2006 Joint Management Speciality Conference*, Salt Lake City, Utah, February 19–22. Paper TUE6A.

Pollard, S.J.T., J.E. Strutt, B.H. MacGillivray, P.D. Hamilton, and S.E. Hrudey. 2004. Risk Analysis and Management in the Water Utility Sector—A Review of Drivers, Tools and Techniques. *Transactions of the Institutions of Chemical Engineers*, 82(B6):453–462.

Pollard, S.J.T., J.E. Strutt, B.H. MacGillivray, J.V. Sharp, S.E. Hrudey, and P.D. Hamilton. 2006b. Risk Management Capabilities—Towards Mindfulness for the International Water Utility Sector. In *Water Contamination Emergencies: Enhancing Our Response*. Edited by K.C. Thompson and J. Gray. Cambridge: Royal Society of Chemistry.

Pollard, S.J.T., R. Yearsley, N. Reynard, I.C. Meadowcroft, R. Duarte-Davidson, and S. Duerden. 2002. Current Directions in the Practice of Environmental Risk Assessment in the United Kingdom. *Environment Science and Technology*, 36(4):530.

Presidential/Congressional Commission on Risk Assessment and Risk Management. 1997. *Framework for Environmental Health Risk Management, Volume I. Risk Assessment and Risk Management in Regulatory Decision-Making, Volume II* [Online]. Final reports. Washington, D.C. Available: <http://www.riskworld.com>. [cited November 14, 2006]

PricewaterhouseCoopers. 2004. *A New Strategy for Success Through Integrated Governance, Risk and Compliance Management* [Online]. A White Paper. Available: <http://www.pwcglobal.com>. [cited November 14, 2006]

Radovanovic, R., and A. Marlin. 2003. Critical Water Mains—The Need for Risk Analysis. *Water*, 30(2):42–46.

Reason, J. 1997. *Managing the Risks of Organizational Accidents*. Burlington, Vt.: Ashgate Publishing.

Rothstein, E., and D. Kiyosaki. 2003. Development of a Strategic Plan: Portfolio Management for Public Utilities. *Journal AWWA*, 95(1):52.

Seddon, J. 2000. *The Case Against ISO9000: How to Create Real Quality in Your Organisation*. Cork, Ireland: Oak Tree Press.

Sharp, J.V., J.E. Strutt, J. Busby, and E. Terry. 2002. Measurement of Organizational Maturity in Designing Safe Offshore Installations. Paper 28421. Oslo, Norway: Offshore Marine and Arctic Engineering.

Solo-Gabriele, H., and S. Neumeister. 1996. U.S. Outbreaks of Cryptosporidiosis. *Journal AWWA*, 88(9):76–86.

Spillett, P. 2003. Environmental Improvements and Capital Maintenance—Have We Got the Balance Right? In *Proc. of the ICE/CIWEM Water Environment Conference: Risk and Reward in Asset Management Delivery—Who Is Best Prepared for the Challenges Ahead?* London: Institution of Civil Engineers/Chartered Institution of Water and Environmental Management.

Stone, S., E.J. Dzuray, D. Meisegeier, A. Dahlborg, and M. Erickson. 2002. *Decision-Support Tools for Predicting the Performance of Water Distribution and Wastewater Collection Systems* [Online]. Cincinnati, Ohio: U.S. Environmental Protection Agency. Available: <http://www.epa.gov>. [cited November 14, 2006]

Strategy Unit. 2002. *Risk: Improving Government's Capability to Handle Risk and Uncertainty* [Online]. London: The Strategy Unit. Available: <http://www.number-10.gov.uk/SU/RISK/ risk/home.html>. [cited November 14, 2006]

Strutt, J.E., B.H. MacGillivray, J.V. Sharp, S.J.T. Pollard, and P.D. Hamilton. 2005. Measuring Risk Management Capabilities. In *Proc. of the AwwaRF International Workshop: Risk Analysis Strategies for Better and More Credible Decision-Making*. Edited by S.J.T. Pollard, S.E. Hrudey, L. Reekie, and P.D. Hamilton. Denver, Colo.: AwwaRF and Cranfield University, U.K.

Strutt, J.E., J.V. Sharp, J. Busby, G. Yates, N. Tourle, N. Fraser, and G. Hughes. 1999. Development of Design Indicators for the Effective Management of Health and Safety During Offshore Design. Paper presented at the Offshore Europe Conference, Aberdeen, September 7–10.

Strutt, J.E., J.V. Sharp, J. Busby, G. Yates, N. Tourle, and G. Hughes. 1998. Development of Design Safety Performance Indicators for Improved Safety Offshore. Paper presented at the ERA Conference on Hazard Management of Offshore Installations, London, December 1.

Swain, H., F. Lazard, and J. Pine. 2005. *Watertight: The Case for Change in Ontario's Water and Wastewater Sector*. Report of the Water Strategy Expert Panel. Toronto, Ont.: Publications Ontario.

Taylor, T. 2005. Achieving a Good Safety Culture—The People Dimension in Health, Safety and Environmental Performance. Presented at Hazard Forum Open Meeting, Westminster, London, March 10.

Toronto Stock Exchange. 1994. *Where Were the Directors? Guidelines for Improved Corporate Governance in Canada*. Toronto, Ont.: Toronto Stock Exchange.

Toronto Stock Exchange. 1999. *Report on Corporate Governance, 1999. Five Years to the Dey* [Online]. Toronto, Ont.: Toronto Stock Exchange. Available: <http://www.otpp.com>. [cited November 14, 2006]

Uff, J. 1996. *Water Supply in Yorkshire—Report of the Independent Commission of Inquiry*. London: Tourcrete.

U.K. Ofwat (U.K. Office of Water Services). 2000. Ofwat letter MD 161. April. Birmingham, U.K.: Ofwat.

U.K. Ofwat (U.K. Office of Water Services). 2003. *Setting Water and Sewerage Price Limits for 2005–2010: Framework and Approach*. Birmingham, U.K.: Ofwat.

UKOOA (U.K. Offshore Operators Association). 1999. *Industry Guidelines on a Framework for Risk Related Decision Support*. London: UKOOA.

UKWIR (U.K. Water Industry Research). 2002. *Capital Maintenance Planning: A Common Framework—Volume 1: Overview*. Report Ref. No. 02/RG/05/03. London: UKWIR.

UKWIR (U.K. Water Industry Research). 2003. *Managing Microbial and Chemical Risks From Source to Tap*. Report 03/DW/02/31. London: UKWIR.

USEPA (U.S. Environmental Protection Agency). 1992. *Framework for Environmental Risk Assessment*. Washington, D.C.: USEPA.

USEPA (U.S. Environmental Protection Agency). 1996. Safe Drinking Water Act Amendments. Washington, D.C.: USEPA.

USEPA (U.S. Environmental Protection Agency). 1998. *Guidelines for Ecological Risk Assessment*. Washington, D.C.: USEPA.

USEPA (U.S. Environmental Protection Agency). 2005a. EPCRA Overview [Online]. Available: <http://yosemite.epa.gov/oswer/ceppoweb.nsf/content/epcraOverview.htm>. [cited November 2005].

USEPA (U.S. Environmental Protection Agency). 2005b. RMP Program Overview [Online]. Available: <http://www.epa.gov/cgi-bin/epaprintonly.cgi>. [cited November 2005].

USEPA (U.S. Environmental Protection Agency). 2005c. Vulnerability Assessments [Online]. Available: <http://cfpub.epa.gov/safewater/watersecurity/home.cfm?program_id=11>. [cited November 14, 2006]

U.S. General Accounting Office. 2004. *Water Infrastructure* [Online]. GAO-04-461. Washington, D.C.: U.S. General Accounting Office. Available: <http://www.gao.gov/new.items/d04461.pdf>. [cited November 14, 2006]

U.S. Government Accountability Office. 2006. *Securing Wastewater Facilities*. Report to Chairman, Committee on Environment and Public Works, U.S. Senate. GAO-06-390. Washington D.C.: U.S. Government Accountability Office.

USNIJ (U.S. National Institute of Justice). 2002. *A Method to Assess the Vulnerability of U.S. Chemical Facilities* [Online]. Washington, D.C.: USNIJ. Available: <http://www.ncjrs.org/pdffiles1/nij/195171.pdf>. [cited November 2005]

Walker, R. 2005. Risk Management—The Realities of Implementation. In *Proc. of the AwwaRF International Workshop: Risk Analysis Strategies for Better and More Credible Decision-Making*. Edited by S.J.T. Pollard, S.E. Hrudey, L. Reekie, and P.D. Hamilton. Denver, Colo.: AwwaRF and Cranfield University, U.K.

Weick, K.E., and K.M. Sutcliffe. 2001. *Managing the Unexpected—Assuring High Performance in an Age of Complexity*. University of Michigan Business School. San Francisco: Jossey-Bass.

Westerhoff, G.P., D. Gale, P.D. Reiter, S.A. Haskins, and J.B. Gilbert. 1998. *The Changing Water Utility: Creative Approaches to Effectiveness and Efficiency*. Denver, Colo.: American Water Works Association.

WHO (World Health Organization). 2004. *Guidelines for Drinking Water Quality* [Online]. 3rd ed. (draft). Available: <http://www.who.int>. [cited November 14, 2006]

Wirth, N., and A.J. Siebert. 2000. Identifying and Evaluating Process Hazards With Process Hazard Analysis. *Pollution Engineering*, 32(13):38–40.

Zsidisin, G.A., A. Panelli, and R. Upton. 2000. Purchasing Organization Involvement in Risk Assessments, Contingency Plans, and Risk Management: An Exploratory Study. *Supply Chain Management*, 5(4):187.

ABBREVIATIONS

AIRMIC	Association of Insurance and Risk Managers
ALARM	National Forum for Risk Management in the Public Sector
ALARP	as low as reasonably practicable (a regulatory approach to risk reduction adopted by the U.K. Health and Safety Executive)
ASCE	American Society of Civil Engineers
AS/NZS	Australian/New Zealand Standards
AWWA	American Water Works Association
AwwaRF	Awwa Research Foundation
BSE	bovine spongiform encephalopathy
CIWEM	Chartered Institution of Water and Environmental Management
CM	change management
CMM	capability maturity model
COSO	Committee of Sponsoring Organizations of the Treadway Commission
CRSA	control risk self-assessment
CSA	Canadian Standards Association
EPCRA	Emergency Planning and Community Right-to-Know Act
ERAC	establishing risk acceptance criteria
ERM	enterprise risk management
E&T	education and training
FMEA	failure mode and effects analysis
FMECA	failure mode and effects criticality analysis
GIS	geographic information system
GRC	governance, risk, and compliance
HACCP	hazard analysis and critical control points
HAZOP	hazard and operability
HS&E	health, safety, and environment
IACCM	International Association for Contract and Commercial Management
ICAEW	Institute of Chartered Accountants in England and Wales
ICE	Institution of Civil Engineers
IEC	International Electrotechnical Commission
ILSI	International Life Sciences Institute
IRM	Institute of Risk Management
ISO	International Organization for Standardization
IT	information technology
IWA	International Water Association

L	liter
LEADA	Leading Edge Asset Decisions Assessment
ML/d	megaliters per day
NAS	National Academy of Sciences
NERAM	Network for Environmental Risk Assessment and Management
NHMRC	National Health and Medical Research Council
Ofwat	Office of Water Services (United Kingdom)
QRA	quantitative risk analysis; quantified risk assessment
RA	risk analysis
R&D	research and development
RBDM	risk-based decision making
RKM	risk knowledge management
RM-CMM	risk management capability maturity model
RMF	risk monitoring and feedback
RMP	risk management plan
RMT	risk management team
RR	risk response
SCRM	supply chain risk management
SERM	Saskatchewan Department of Environment and Resource Management
SRP	strategic risk planning
UKOOA	U.K. Offshore Operators Association
UKWIR	U.K. Water Industry Research
U.K.	United Kingdom
U.S.	United States
USEPA	U.S. Environmental Protection Agency
USNIJ	U.S. National Institute of Justice
WEF	Water Environment Federation
WHO	World Health Organization

IWA Publishing's authorised EU representative for General Product Safety Regulations is Diane D'Arras, 15 rue Duret, 75116 Paris, France, e-mail: safety@iwap.co.uk.

Printed and bound by CPI Group (UK) Ltd, Croydon, CR0 4YY
17/09/2026
02218802-0012